Sherman F.

About the Cover

The C-130 on the front cover is a B-model. It belongs to the 145th Airlift Group at Charlotte, North Carolina.

It's Air National Guard crew is banking in salute to the Wright memorial at Kitty Hawk, North Carolina.

The 145th's call sign is "Epic" ... as in the epic of flight.

Listen for us on the radio.

SHERMAN F. MORGAN

Old Planes, Young Men and Red Wooden Shoes

For my Oshkosh
pal Bob — N 8369 W
captain! Thanks
for the story.

Sherm F Morgan

7/29/94
OSH

Pendragon Publishing Co.
1484 Old Tara Lane
Fort Mill, S.C. 29715

1

First Edition, December 1993

Printed and bound in the United States of America.

Library of Congress Catalog Card Number:
93-84267

International Standard Book Number:
0-944792-26-X

Also by Sherman F. Morgan

The Aviation Humor of 1987

The Little Apple Orchard

Good Sticks

Classic Aviation Humor—Book II

Putu, Siku & Kanik

Classic Aviation Humor—Book III

Acknowledgements

Thanks are again in order for all of the fine folks who sent their funny stories to me for this book—especially Larry Struck, who sent more than anyone. I tried to give everyone credit for their stories as I used them.

Thanks also to Ed Dingivan, who serves as my proofreader and quality control barometer for these stories. Ed is a perfect specimen of the target audience that I'm writing for. If Ed likes the book when he proofreads it, I know that it's chances of success in the marketplace are excellent.

Thanks also to LTC Pick Freeman for his fine assistance in proofreading—and for keeping us out of jail during the "research" sessions.

Special thanks to my friend Stephen Coonts for his permission to borrow from his excellent title, *The Cannibal Queen.*

Be thinking of a good story to tell me when you drop by the author's corner exhibit at Oshkosh this year. See you there.

Dedication

This one is dedicated to a couple of wonderful ladies who have looked after my publishing endeavors since 1990—Wanda and Vanessa Phelps. Words alone can not express my gratitude for their faith, good humor and hard work.

I know that they won't mind sharing the dedication with our mutual friend, Todd Branson.

I'm sorry I didn't get this one out in time for you to read it Todd, but I can hear you laughing at it just the same.

Foreword

The most common question interviewers ask me is, "Are all of your stories true?"

I always answer by saying, "I have no idea whether they're true or not—I just know a funny story when I hear it."

That's why I always put that disclaimer in the front of my books, saying that any resemblance to actual events, locales or persons is entirely coincidental. I make no effort whatsoever to confirm any of the facts in the stories that are related to me by voice or mail.

Since I don't know what portion of the stories is true and what's not, I've always presented these stories as pure fiction, although of course most good fiction has some basis in fact. The truth is usually stranger than fiction—that's because fiction has to make sense.

In the case of this book, most of the book is completely true. The great majority of the characters in the story are real, I used their real names. The places, times, and events are all as accurate as I could determine them to be. Hard facts such as aircraft tail numbers, flight times, flight routes, crew compositions and dates are also as accurate as I could determine them to be.

So, where did I take literary license?

In the dialogue.

Some of the conversations that I included in this book were pure fiction. Some of the dialogue never took place at all, while others occurred in different settings or at different times than I reported.

Why change or invent dialogue? Two reasons: to enhance the entertainment value of the book, and to efficiently advance the story line.

By "enhancing the entertainment value", I mean that I might have included a funny line in a story that actual wasn't said when the story was originally told. If I thought that including the new line would make the story more humorous to the reader, I stuck it in.

By "efficiently advancing the story line", I mean that I may have rearranged dialogue that actually occurred on three different flights, and used it all on one flight. I didn't write down everything that happened during every second of the deployment—sometimes I lumped conversations together.

Also, I sometimes use dialogue as transitions between chapters or story line sequences. To make that happen, I sometimes have to invent or rearrange dialogue.

I also wrote this book months after the fact. I wrote strictly from memory, without detailed notes. While it's easy enough to go back and check on how long it took to fly some place, it's nearly impossible to come up with a consensus on who said what to whom where.

Why highlight these inaccuracies in the dialogue?

Because I don't want any of my friends in any of these stories to be offended by inaccuracies in the dialogue. But, I also don't want to call the whole book a piece of fiction, because it's not.

There is a very large community of part time military personnel in the Reserve and Guard, who will readily identify with all of the characters, frustrations, events and adventures that I describe in this book.

A large percentage of the pilots in the Reserve and Guard are also airline pilots, so a lot of the humor in this book is derived from airline war-stories. These stories are told in the cockpit and BOQ rooms in this book, just like they are in the real world.

That's the real purpose of this book, to shed a little insight into what it's like to go on a deployment with an Air National Guard squadron. The dialogue is mostly used to relate the spirit of camaraderie between the members of the squadron.

I don't think anything that I've included in this book in the way of dialogue inaccuracies will detract from the overall picture that I've tried to sketch for you of what it's like to be on a deployment. My hope is that if you are a weekend warrior that you will laugh at the stories and say to yourself, "My unit is *exactly* like that."

If you're not in the military, I hope you will laugh heartily at the stories and gain some new insights into what that life style is all about.

On the serious side, the chapters about my Uncle Eugene and Margraten contain no inaccuracies that I am aware of.

As always, write to me c/o Pendragon Publishing Co. if you have a funny aviation story for my next book, or if you feel like telling me what you thought of this one.

Chapter 1

Arthur "Eugene" Robertson was born on January the second, 1924.

Completely unaware of the poverty of the rural Missouri farm where his cradle lay, Eugene was only cognizant of the unbridled love showered upon him by his twenty-three year old father, Roy, and especially his sixteen year old mother, Irene Jane.

As often happens when young girls bring babies into the world, people were soon saying that Irene and Eugene were growing up together.

With Roy forced away from home a great deal of the time earning the young family's living, mother and son were left alone to forge a bond which would eventually prove to be stronger even than death.

Lithe, intelligent, athletic, fiercely competitive— Eugene exhibited all the characteristics of the quintessential first-born.

Eventually six more siblings would be born into the family for Eugene to help raise—three brothers, and three sisters. The youngest two sisters, Mary Imogene and Estelene, were twins.

One form of entertainment the younger children especially enjoyed were Eugene's acrobatic shows. After warming up with a series of handstands and backflips, he would string a wire between the roof of the house and the eave of the barn, and walk the wire carrying a cane fishing pole to assist his balance.

So far as anyone could remember Eugene never lost a fight, though it took an extraordinary provocation to goad him into one—such as a bully harassing one of the younger children. He was unashamed of his momma's-boy reputation, he could be tough as nails when he needed to be.

Irene's constant attentions nurtured an especially sensitive nature in Eugene which no one but Irene fully appreciated, until the day Eugene's dog, Roadie, was accidentally killed by a passing car.

The grief-stricken boy was completely lost—unable to fathom a way to deal with the loss of his beloved dog, which he had raised from a puppy.

Finally, Irene urged Eugene to draw a picture of Roadie and write down how he felt about the dog, "So other people will always know and remember Roadie."

It was the first exhibition of Eugene's natural God-given talents for art and verse. His pencil rendering of Roadie was remarkable, and the short heart-felt poem he penned below the picture rivaled any verse in the Book of Psalm.

Eugene's youngest sisters, the twins (my mother Mary Imogene and my aunt Estelene) often said that their most vivid memories of Eugene as a boy were his homecomings. The cedar-lined lane that led to their home ran east-west.

Eugene would walk home in the evenings after his work was done, with the setting sun behind him silhouetting his lanky frame. Long before the silhouette was close enough to recognize the features, my mother and aunt Estelene would know it was Eugene, because he was always carrying a notebook in the crook of his left arm, penning with his right. Their cries announcing his arrival never failed to bring Irene to the front porch, with her face beaming brighter than the dying sun.

At nights, Irene embellished Eugene's education with a special gift beyond the means of his one-room school; she taught him Latin.

While he was still an adolescent, Eugene's natural leadership abilities were noticed by a wealthy local farmer, Mr. Bell. Mr. Bell hired Eugene to come to his large farm and help him manage it, especially with the keeping of his books.

In short order, Eugene fell hopelessly in love with Mr. Bell's daughter, Theda, and, against his mother's wishes, married her when he was sixteen.

Irene felt that in spite of Eugene's mature character, that he was too young to marry. She begged him to go and see the world first.

But, it was 1940, and Europe was in flames. Eugene intuitively knew that he had better live and love fast, for the world he was destined to see was menacing.

Eugene and Theda had one daughter, Jeanie, before the Japanese attacked Pearl Harbor and plunged the United States headlong into World War II.

Eugene immediately joined the Army, where his

inherent leadership skills were put to the ultimate test as a combat infantry soldier for the US Army's 84th Division.

Theda and Irene received countless letters from Eugene, chronicling the heroics of his unit as they fought their way across Europe. Some of the letters included black-and-white photographs of Eugene and his friends, and the war-ravaged people of western Europe that they were liberating.

Two Dutch sisters appeared with Eugene in several of the photographs he sent home while his unit was fighting its way across Holland. The sisters had been crippled during the German occupation.

The young Dutch ladies made their living sewing dresses in a small shop that had been wrecked during the retaking of their town. The small cart that they used for transportation had also been nearly destroyed.

Eugene's unit received a few days rest after retaking the sisters' village. Eugene used the time off to help the sisters repair their wagon and their dress shop. He gave them his mother's address and invited them to write to her.

They did, in the only common language they knew, Latin. Irene was so pleased to hear from them and the gratitude that they felt toward her son, that she sent them a Sears catalog and invited them to write and tell her anything in the catalog that they would like for her to send to them.

They asked for silk scarves and stockings. Irene sent them, and in return they mailed wooden shoes and Dutch chocolates back to Missouri.

Eugene's second daughter was born while he was in Europe, recovering from wounds received from a German sniper. He wrote home with specific instructions, that his daughter be named Lilma Iris. He was adamant that she should always be referred to by her proper name, Lilma, and never nicknames such as Libby. He said he would explain the name when he returned home to see her for the first time.

Eugene's wounds were serious enough to justify his return to the States, but he insisted on staying with his unit. His Marine younger brother, A.J., home after being seriously wounded on Okinawa, wrote to him begging him to come home.

Ever the fierce competitor, Eugene wrote back that he had never been in a fight that he failed to finish. Besides, he had new men under him that would stand a better chance of seeing the end of the war if he was there to lead them.

On April 4th, 1945, just 34 days before the war ended, Eugene was leading a squad of men on an armed reconnoiter of a road bend. The commander of the Sherman tanks that Eugene was scouting for, wanted to make sure that the Germans didn't have a tank trap set up on the other side of the hairpin turn before he exposed his tanks.

Eugene spotted the camouflaged 88 mm gun just

before it fired. He shouted the warning to his squad just as the canon fired, barely giving his men enough time to dive for cover before the exploding shell laced the air with deadly shrapnel. Firing his rifle as he fell, Eugene provided what covering fire he could while his men scrambled back around the bend.

By the time the Sherman tanks destroyed the now-exposed German anti-tank gun, Eugene had lost too much blood to survive.

He was buried along with more than 32,000 other fallen Americans at the hastily conceived Margraten American Military Cemetery, close to Maastrich, Holland.

Soon after V-E Day (May 8, 1945), the U.S. Army contacted the family about their wishes for Eugene's body. Irene was very ill at this time, but Roy desperately wanted his son brought home. Finally, Roy decided not to bring Eugene home to the family cemetery at that time, for fear that the shock would be too much for Irene.

For the next four years, each Memorial Day, the crippled Dutch sisters that Eugene had befriended would go to Margraaten and place flowers on his grave. They mailed pictures of the flower-bedecked wooden cross with his name on it, home to Roy and Irene.

In 1949, Roy received a letter from the Army stating that the cemetery at Margraaten was about to be reconfigured. Roy was afraid that if he didn't have

Eugene's remains sent home then, that he would never be able to.

Because she was so gravely ill, the family tried to keep Eugene's reinterment in the family plot a secret from Irene. She knew, without being told. As soon as Roy came to her bedside, she said, "You've brought Eugene home, haven't you."

She smiled, pleased with his answer. She died 30 days later, May 10th, 1949, on Mother's Day. She was buried in the Black Creek cemetery in Poplar Bluff, Missouri, next to Eugene.

Twenty-six years later I took emergency leave from Army flight school at Fort Rucker, Alabama to be a pall-bearer at my Grandpa Roy's funeral. My brother, Mike, was a Marine then at Camp Pendleton, California. Mike took emergency leave to be a pall-bearer also.

Mike and I wore our military dress greens as we laid Grandpa Roy down on Irene's left, with Eugene on her right.

As I was growing up, my mother and my uncles and aunts on the Robertson side of the family were always fond of reminiscing about Eugene. I almost felt like I knew him. Being the first born of my family and interested in competitive sports, good books and writing, I always felt like I had a lot in common with Eugene.

As a child, I remember being intrigued with the story of Eugene's double burial. There was something fascinating about lying with comrades in a foreign soil that he helped to make free, and then returning home.

I remember trying to talk to Grandpa Roy about it once when I was a little boy. This was probably twenty years after Eugene's death. Perhaps it was the innocence in the way I asked the questions, but it was the only time that I ever remember seeing my grandpa cry. His grief upset me so much that I never spoke to Grandpa about Eugene again.

I did, however, continue to harbor a fascination with the Margraaten cemetery. I've always wanted to go there and see it for myself.

As luck would have it, I was sitting about eight rows back in the auditorium of my Air National Guard squadron last March, while Dave Hatley was trying to put together a workable flying schedule for the month.

Dave is the chief pilot scheduler for our C-130 unit, and those pilots sitting about eight rows back are his chief hecklers, responsible for drawing out as many laughs as possible from whatever serious effort Dave is engaged in.

Almost as an afterthought, Dave mentioned that we might want to think about freeing up some time in May if we could, since we were going to be sending four crews on a deployment to Belgium for the first three weeks of May.

"Belgium!" I said outloud, "That's right next door to Holland!"

"Very good Sherm," Joe Sepko said behind me, "maybe they'll let you go if you can tell us its capital."

"Brusselles?" I ventured.

"Sprouts or Belgium?" Joe quipped back.

Dave cut through the chuckles with, "Looks like we have our first AC (aircraft commander)."

And with that, I found myself embarked on my long-awaited journey to visit Eugene's original resting site, at Margraaten, Holland.

Chapter 2

My inclusion as an aircraft commander on the Belgium deployment did not occur quite as casually as I described it in the previous chapter.

It started that casually, but the squadron planners went through a fairly intense selection process before finally deciding exactly which crewmembers would be allowed to go. There were far more volunteers than available slots.

I asked one of the chief schedulers, "Junkyard", exactly what criteria they used to make the final selections.

Junkyard told me, "We used the same criteria we always use when selecting players—we look at the most likely missions to be performed, then we evaluate the candidates' qualifications and currency, and their availability... then we pick the one with the big tits."

Since my name was the first one in the hat, I was one of the lucky ones selected to participate in exercise "Volant Partner '92".

Volant Partner '92 was a joint exercise between the Belgian forces stationed at Koksijde Air Base, on the coast of Belgium, and the forces of several C-130-equipped Air National Guard units from around the United States.

Besides my own unit, the exercise included crews and aircraft from Guard units based in Kentucky,

Maryland, and Delaware.

My guard unit, the 68th Airlift Squadron of Charlotte, North Carolina, provided three C-130B aircraft and four crews for the exercise. We also deployed more maintenance personnel than anyone else, which probably explains why we wound up with the best maintenance record of the exercise, in spite of being equipped with the oldest airplanes.

Since we were deploying four crews on three aircraft, one of the crews had to be broken up for the flight from Charlotte to Koksijde. LTC (Lieutenant Colonel) Pick Freeman's crew wound up being the odd-crew out for the flight over. Pick flew with Todd Kelly and Ray Byrum in the last aircraft. Pick's copilot, James Talbert, flew with me in the second ship.

Believe it or not, this simple crew shuffling led to the first adventure story of the deployment.

My crew's show time was 0930 on May 1st, 1992. The first crew, commanded by Artis Galbreath, showed 30 minutes before mine. This allowed them time to place their bags in the designated vehicles and report to the squadron's auditorium for their departure briefing.

The third crew, commanded by Todd Kelly, showed 30 minutes after us.

Everything went like clockwork. Everyone's bags wound up on the correct aircraft, and Artis's crew finished their departure briefing just as my crew showed up at the auditorium to receive our briefing.

The detachment commander, LTC Gary Wilfong, conducted the departure briefings. The plan was for Gary to brief all the crews, and then jump on the first aircraft to depart. That way he would always be with the first crew to arrive anywhere, so he could help solve all the problems that inevitably pop up, such as billeting, parking, fueling, etc.

Now remember that James Talbert, ("JT"), the copilot of the fourth crew, was riding with me. Traditionally, the copilot is responsible for insuring that his crew's "trip kit" makes it onboard the airplane.

The trip kit is a large briefcase which includes customs documents, various report forms, and such invaluable items as passports and shot records for the crew. It is very important that the trip kit make it onboard the airplane, with all of the necessary documents inside.

Being a good copilot, JT checked the contents of his crews' trip kit. Then, since he didn't have an aircraft of his own, he made sure that his crews' trip kit got put onto my airplane.

The problem arose later, when JT's aircraft commander, Pick Freeman, decided to take his passport out of the trip kit and carry it on his person.

Pick didn't think about this until I was already taxiing out for takeoff. Pick was still in the operations building, so he had no way of checking that JT had taken care of their trip kit. Pick was especially concerned about checking that his passport was actually in the trip kit.

Bedlam reigned for a short while as the operations people tried to determine exactly what had happened to Pick's trip kit. Messages got crossed and ideas confused, until somebody realized that JT was on my airplane, and that he probably had the trip kit with him.

We were climbing out over South Carolina by that time, well on our way to our first fuel stop in Bermuda, when departure control called us and forwarded a message from our operations asking us to call home concerning Pick's passport.

It just so happened that we were flying aircraft 61-02636, a bird notorious for its poor UHF transmitter. We tried to contact "Newsreel" (our squadron operation's callsign), but they were unable to read our UHF transmissions. Newsreel has no VHF capability.

Newsreel was able to discern that we were trying to call them, so they transmitted in the blind that they needed to know if we had Pick's passport on board our aircraft.

JT confirmed that we did have the passport on-board, and my copilot, Dave Richards, thought of a way to get that information to them.

As this book goes on you're certainly going to

learn a lot more about the eight pilots that accompanied me on this trip, especially Dave Richards, but for now let it suffice to say that Dave is rarely at a loss for ideas whenever a problem arises.

Dave proposed that we ask departure control for permission to leave their VHF frequency for a minute, which they granted. With our only VHF radio now free, we called back to Charlotte ground control, and asked them to use their direct phone line to call Newsreel and forward the information that we had Pick's passport onboard.

Ground control forwarded the message for us, and the last crew was finally able to depart, assured that Pick would have a passport waiting for him when he arrived in Bermuda.

We switched our VHF radio back to departure control, and soon settled in for the 3.3 hour trip to Bermuda.

It wasn't long before we were coasting out over Charleston, South Carolina. As soon as we switched to the oceanic control radio frequencies the radio chatter calmed down to just a few intermittent calls. These greater periods of silence on the radio always encourage more chatter on the intercom, and that's exactly what happened in our cockpit.

JT was flying copilot for me in the right seat. Normally Dave would have been there, but JT is still building flight time so that he can get his airline job. Dave is already a captain for USAir, so he gladly allowed JT to log the 15.1 hours of flight time it took us to fly from Charlotte to Koksijde.

Dave is also an aircraft commander in the C-130, but at the time we started this mission he didn't have a lot of experience flying in Europe. So, he volunteered to go along on this deployment as a copilot.

Dave was lying on the top bunk at the rear of the cockpit, reading a magazine, when JT twisted around in his seat and said, "Dave," over the intercom.

Dave glanced forward and saw JT twisted around looking at him, so Dave thumbed his intercom switch and said, "Yea buddy, you tired already?"

JT laughed and then said, "No man, I'm just getting started, but I wanted to ask you something about Pick."

"What about him."

"Doesn't he have a cabin up in the mountains?"

Dave immediately started laughing so loud that we could hear him even without his intercom button depressed.

After a minute or so, when he calmed down enough to talk, Dave said, "Have you heard the story about Pick cutting down the pine tree?"

JT adjusted his sunglasses and said, "I just heard that you had a good story about Pick and his cabin, but I didn't hear exactly what it was about."

By now all of our curiosities were piqued. Dave put down his magazine and raised himself up on his left elbow on the bunk, and for the next several minutes nobody said a word while Dave told us the story of Pick's lumberjack adventure.

As closely as I can recall, this is the way Dave told the story, in his voice:

Mark Frederickson and Pick and I flew a night local training mission together a few weeks ago. We went over to Pope (Air Force Base) to log some night-assault landings. On the way home from Pope, Pick started telling us about this cabin that he bought up in the mountains of western North Carolina. The cabin is just off the Blue Ridge Parkway, in a neat little community called Little Switzerland.

Pick kept talking about how beautiful it was up there in the spring. He got Mark and me so worked up about it, that we finally wrangled an invitation from him to go up there and spend the weekend with him. Little did we know, that Pick had an ulterior motive in mind.

When we showed up that Saturday, Pick showed us around his place. It really was a neat place, nestled up there in the mountains with a great view. But, it didn't take us long to figure out what Pick had in mind.

As we examined the backyard, Pick was careful to point out the sixty-foot spruce pine on the edge of his property, which was leaning towards his neighbor's cabin.

Pick took a sip of his beer and said, "Yep, I'm

going to have to cut that darn tree down before a strong wind blows it over into my neighbor's cabin and kills somebody."

Quick as a wink, Mark said, "Well heck Pick, let's go find a chainsaw and Dave and I will help you buzz that rascal up."

Before I could think of a way to retract Mark's offer, Pick pointed to a storage area on the back of his cabin and said, "My saw and axe are in the tool crib, do you feel up to doing it today?"

What could we say? I could tell by looking at the tree that we were probably getting into something that we shouldn't be messing with. But, the snowball had already started tumbling down the mountain—it was too late to stop it without getting buried in the avalanche.

I said, "Pick, how are we going to keep the tree from falling on your neighbor's house when we cut it down?"

He said, "We'll just top it."

"Top it?"

"Sure, we'll just cut it down ten feet at a time, starting from the top. That way there will never be enough of it falling at one time to hit the cabins."

Mark and I must have both been wearing fairly pained expressions, because Pick immediately added, "Don't worry, I'll do the topping. I just need you two to pile up the limbs for me as I cut them so I don't wind up with a huge mess at the base of the tree when I get down to the bottom."

Mark and I grinned at each other and said, "We can do that."

Actually, I think both of us were starting to enjoy the idea of watching what would surely be quite a spectacle—Pick climbing a tree with a chainsaw.

Pick drained his beer and then headed for his tool crib. He soon emerged, carrying a rusty motor with a bar sticking out it, a baseball bat with some metal on the end, and two plastic jugs.

It didn't take too long to figure out that the rusty motor was supposed to be a chainsaw, but the bat was what captivated me.

I picked up the bat while Pick unscrewed the lid to one of the plastic jugs. I think the jug had originally contained fabric softener, but Pick had apparently refilled it with gasoline.

While Pick poured the gasoline into the chainsaw, I asked, "Where did you get this Pick?"

Without looking up he answered, "I found the axe head under the porch. I didn't have a handle for it, so I just whittled down the top of that Louisville slugger until the head would fit on it."

The second plastic container held recycled motor oil. Pick busied himself filling the chainsaw's chain-lubricating reservoir while Mark and I admired the bat-axe.

Mark said, "Do you take a level swing with this thing Pick, or is it okay to chop down on the ball?"

Unruffled, Pick screwed the oil-reservoir cap back on the chainsaw as he answered, "It doesn't matter how you swing it boy, what matters is that the price was right."

Mark said, "You really are an airline captain, aren't you Pick."

Pick grinned broadly and said, "What was your first clue?"

Pick turned the chainsaw right-side-up again and said, "Stand back boys, this thing hasn't been started since hurricane Hugo came through—I'm not sure what she's going to do when I pull on this rope."

Mark and I took a wary step backward and watched while Pick flipped on the ignition switch, adjusted the choke, and gave the starting rope a mighty pull. The motor turned over with a low-pitched gurgle, and quickly stopped.

Pick jerked on the rope again, and the motor answered with a few heartless grunts before falling silent once more.

Pick adjusted the choke and said, "The gas is a couple of years old—it might have gummed up the carburetor."

Mark said, "Do you think so, after just sitting for a couple of years?"

I looked him in the eye and said, "You're about ready to check out as captain yourself, aren't you Mark?"

Pick gave the motor about a dozen pulls without

any results that could be described as promising. Once a puff of white smoke spewed out of the muffler, which Pick said indicated that the plug was firing. Mark and I thought it probably indicated that the saw was about to catch on fire.

While Pick took a short breather to rest his arm, Mark said, "Did you hear that story about the Boeing flight test crew up in Seattle that was taxiing out behind a Northwest Airlines crew in an Airbus?"

Probably just to give his arm a little longer rest, Pick shook his head, so Mark continued, "When the Boeing crew called ground control for taxi clearance, the controller said, 'Just follow the Northwest airplane to runway 31.'

"The Boeing crew answered, 'Roger, we'll follow the Homelite.'

"The Airbus pilot squeezed his mike button and said, 'Doesn't that Boeing pilot know the difference between an Airbus and a chainsaw?'

"Before the ground controller could say anything, the Boeing pilot responded, 'Everyone knows the difference—about 60,000 trees per hour!"

It was probably Pick's laughter that finally roused Pick's neighbor out of his house. His neighbor was about eighty years old. He had always lived in the North Carolina mountains. His name was Rufus.

Rufus waddled across his yard toward us, with a huge straggly mutt at his side.

When he was close enough to hear our answer,

Rufus shouted, "Hey Pick."

Pick answered, "Hey Rufus, how are you doing?"

"Fine Pick, just fine," he answered while he eyed Mark and me and the brews in our hands.

Pick used a little more volume than normal as he introduced us, "These are friends of mine Rufus, they're here to help me cut down that tree that's leaning toward your house. This is Dave Richards and Mark Frederickson."

Rufus looked both of us in the eye as he shook our hands.

While Mark shook Rufus's hand he asked, "What's your dog's name?"

"Boone," he practically shouted back at Mark.

Mark gave him a quizzical look, so Rufus explained further, "This dog hates bears, he tries to kill everyone that comes by the place. Daniel Boone supposedly killed his first bear when he was only three. That's why I call him Boone."

Mark got a little wide eyed—about like a young owl looking at you for the first time. Then he asked, "Where do the bears hang out when they're not around the house?"

Rufus pointed up the hill and said, "The beehives. They're always coming around the beehives out in front of the house. They like to knock the top off the hives and steal the honey, but ol' Boone here keeps them away."

Pick interjected, "The beehives are in that apple

grove in front of the cabins, they're a good fifty yards away. The bears usually don't come any closer than that."

Mark's nervous laugh indicated that fifty yards was slightly less clearance than he normally liked from bears.

Rufus gave Pick's chainsaw a worried glance, then offered, "That's one of the sorriest excuses for a chainsaw I've ever seen. What do you expect to accomplish with that relic Pick?"

Pick ignored Mark's and my chuckles as he waved toward the tree and said, "I'm aiming to cut down that spruce Rufus, but I can't get my saw started."

Rufus made no attempt to hide his laughter, or his obvious disdain for Pick's saw. Finally he said, "Hell, you ain't gonna cut nothing down with that relic Pick. You wait here and I'll be right back with a real man's saw."

Pick tried to stare us down, but Mark and I treated ourselves to a nice long guffaw while Rufus waddled off toward his cabin. Boone trotted off toward the apple orchard.

A few minutes later Rufus reappeared carrying what appeared to be an ancient chainsaw. It must have been manufactured in the 1920's or 30's at the latest.

Rufus set the saw down in front of Pick and said, "Now this baby will get the job done for you."

Pick said, "I've never seen a model like that

before."

"Quit making them before you were born; they lasted too long. Manufacturers don't want to build something that will last forever, they want us to have to buy a new one every few years."

Pick nodded agreeably, then said, "Anything special I need to know about operating it?"

"Ain't no cut-off switch on this saw," Rufus said, "you have to put the choke full-on to stop the motor. New saws all have kill switches, but they don't need them—all they need is a good choke. Manufacturers just put those kill switches on to be able to charge more money."

Mark and I stepped well back out of the way as Pick prepared to start Rufus's chainsaw. Both of us were thinking that it belonged in a logging museum somewhere, but Pick was obviously more-than-willing to try it on the spruce.

Pick tightened his grip on the starting handle, and Rufus reminded him, "Just remember, you have to choke it to kill it."

Pick nodded and gave the start rope and sharp tug. The relic fired and immediately caught, kicking out a cloud of blue smoke. Pick gunned the throttle a couple of times, then let the motor warm up enough so that it would idle easily with the choke completely off.

Satisfied that the saw was ready to go, Pick flipped the choke lever full up and waited for the motor to die. When the motor stopped and it was finally quiet enough to talk again, Rufus asked, "What do you think of that Pick?"

Pick winked and said, "That dog will hunt."

Pick walked over to his back porch where he could sit down and strap on his climbing spikes. As he tightened the metal spikes to the insides of his boots, Pick explained his plan to us.

"I'll climb the tree using these spikes and my safety belt fastened around the tree. I'll use Rufus's chainsaw to cut off the limbs flush with the trunk so my safety belt will slide up the tree as I climb.

"I'll be counting on you two to pick up the limbs as they fall and stack them up over on the downhill corner of my lot, away from Rufus's cabin. I'll burn those limbs this fall.

"When I finally work my way up high enough, I'll cut the top ten feet off the tree. Be careful and don't let it hit you when it falls.

"I'll give you two time to drag that ten-foot section over to the brush pile, then I'll climb down and cut off another ten-foot section. We'll just keep going like that until I work my down to the ground."

Rufus said, "Just be careful not to cut off so much that it falls on my cabin."

Pick nodded as he strapped on his safety belt and said, "I'll be careful of your cabin Rufus, but I'll also have to be careful not to climb too high. These spruce trees are awful limber."

Rufus said, "That's true—that's why they used to make airplane propellers out of them. Just remember that you have to choke that saw to cut it off."

Pick grabbed the saw and said, "I'll remember. Are you two ready?"

Mark and I guzzled the last of our brews and tossed the empty cans onto Pick's back porch. Mark said, "Let's do it."

Pick buried the spikes in the soft bark of the spruce, then leaned back against his safety belt to test it. Satisfied, he gave the start rope a quick pull, and the motor roared into life.

The cutting blade on that old saw was sharp as a razor. Pick barely had to brush the spinning chain against the limbs to make them fall free in a cloud of fresh sawdust.

Mark and I took turns grabbing the limbs as they fell and carried them to the rapidly-growing brush pile. Pick was shuffling his way up the tree so fast that we could barely keep up with him. Rufus stood well back out of the way, apparently enjoying the spectacle.

The tree started swaying more and more as Pick got closer to its top. Convinced that he had climbed as high as he safely could, Pick turned the saw horizontal and held it just above his head to slice through the narrow trunk.

The top ten feet fell away from Pick, landing well short of Rufus's cabin.

I waited until the chainsaw's motor was at idle again so that Mark could hear me, then I said, "This is going a lot better than I expected it to."

Mark said, "Did you hear that?"

"What?"

Mark gave me a puzzled look and said, "Did you hear a yell or something?"

I was just about to answer, "No", when I suddenly thought that I did hear something. It almost sounded like a howl.

A moment later, I was certain I'd heard Mark's mystery noise. Our heads simultaneously snapped around to look toward the apple orchard, where we could now clearly hear the sounds of loud barking and howls. They were growing louder.

A second later, we spotted the source of the noise. A huge black bear was scrambling out of the orchard at a dead run, followed closely by Rufus's dog Boone.

The dog and bear were both running downhill as fast as their legs would carry them. Boone was snapping at the bear's hindquarters as he ran.

The bear was obviously looking for a place to hide, preferably someplace high and dry. Scanning his escape route as he ran, the bear quickly spotted the nice slick-trunked spruce and made for it.

The first indication Pick had of the new situation, was the sudden appearance of the snarling black bear on the other side of the trunk from him! He hadn't heard the earlier going-ons because of the noise of the chainsaw.

With honey-flecked froth dripping from his mouth, the bear quickly decided that he didn't care for Pick's presence on the other side of the trunk. The bear reached back to take a swipe at Pick, but quickly had to check his swing and grab for the trunk again as the tree started to lean over under his added weight.

Pick gunned the motor of the saw and started to started to slice at the bear, but then he had to brace the base of the saw against the tree's trunk to keep from being thrown off as the tree sagged toward the bear's greater weight.

Being a limber spruce, the supple trunk bent almost far enough for the bear's back side to touch the ground, before springing back. The mountainside was a crescendo of sound, with Boone barking and jumping at the bear, the bear growling and frothing and Pick, and Pick yelling and gunning the saw!

With the added momentum built up from its upward swing, the spruce quickly swung through the vertical and continued over until Pick's backside was nearly brushing the ground.

In all the confusion poor Boone must have gotten his targets mixed up. The bear was swiping down past Pick's head, so Boone apparently decided to latch onto the closest target he could reach.

Unfortunately for Pick, Boone's closest target happened to be Pick's hind end!

With Pick's screams now drowning out even the

chainsaw, the tree sprang back again, this time with Boone along for the ride also!

The spruce had built up so much momentum by now that we could barely see it moving upward. The whole scene was a blur, like a ball being swung on the end of a string.

The spikes must have come out just about the time that the tree swung through the vertical. The spikes are designed to overcome the downward pull of gravity—they're not built to resist the upward pull of centrifugal force.

As soon as the spikes let go, Pick's safety belt easily slid up and over the end of the trunk, helping to dislodge the bear's front claws in the process.

With nothing but its back claws dug into the bark, the bear lifted off the tree at the same time Pick did, resulting in a perfect formation. The bear was flying along on his back, with Pick screaming and gunning the chainsaw just above him, and Boone hanging in there in close trail. I'd never seen anything like it.

They disappeared for a few seconds over the apple trees, then we heard the unmistakable sound of crashing wood!

Rufus answered our puzzled looks with a single word... "Beehives!"

About two seconds later, we heard the most amazing cacophony of noises I've ever heard. Barks, growls, howls, yells, screams, curses, and buzzes... every sound you can imagine was coming from that orchard.

Then, we saw them, hurtling pell-mell down the hill. Boone was in the lead, a patch of Pick's trousers still hanging from a fang. The bear was running a close second, making a sound that resembled a cross between a steam whistle and a wounded buffalo.

Pick was close behind. He probably would have been leading, but he had to high-step to run with the climbing spikes on. That and the fact that his climbing safety belt as still swinging from his waist, forcing him to leap over it every fifth or sixth step like a hurdler on a track.

He was still holding onto the chainsaw, gunning the engine and swiping at the cloud of bees that engulfed the entire entourage.

All three of them raced past us at full tilt, straight for the brush pile that Mark and I had so carefully built on the corner of Pick's lot. They hit the pile going so fast that branches were thrown twenty feet in the air.

The last I saw of Pick, was his bare behind racing away below the curtain of falling branches, still gunning the chainsaw and swiping at the bees.

We looked at Rufus for some idea as to how we could help, but all he said was, "He'll have to choke that saw to get it to cut off."

By the time Dave finished his story I could see the enchanted island of Bermuda lying straight off the nose.

Of course, none of us believed a word of Dave's story. That's what made it so funny—it was completely ludicrous.

Pick is famous for his know-how, common sense, and especially for the way that he takes care of his stuff. *Nobody* takes care of his possessions the way Pick Freeman does. The man does periodic maintenance on his garden hoses!

Unbelievable as it was, the story still got the trip off on the right foot. We had a hard time getting everybody to stop laughing long enough to run the checklists.

JT managed to compose himself well enough to make the first landing of the trip. It was a squeaker.

Chapter 3

We parked beside Artis's airplane in front of base operations at NAS (Naval Air Station) Bermuda.

After briefing the ground-handling people that we just wanted a quick gas-and-go, Dave Richards and James Talbert and our navigator, Terry Heath, and I started walking across the ramp to base ops.

We met Artis, Brendan and Gary about halfway across the ramp. They were finished flight planning, so they were walking back to their airplane. They planned to takeoff as soon as they got their gas on board.

Everyone was in high spirits. The 3.3-hour flight from Charlotte hadn't really been enough to tire everyone out yet, (especially in my airplane with Dave telling his stories), and the next leg to Lajes, Azores didn't seem too intimidating.

"Where do you guys want to rendezvous in Lajes?" was the first question out of Dave's mouth as the two crews faced off on the hot ramp.

Gary said, "It will be two o'clock in the morning when we get there—I don't know if anything will be open."

Before we left Charlotte, I had told Dave about a trip I had in C-141s years ago out of Travis AFB, when my crew had stayed up all night drinking cans of Heinekens in the Lajes bowling alley.

Dave immediately volunteered, "Sherm says the

bowling alley is open twenty-four hours a day, and they have Heinekens in cans!"

Artis smiled and said, "The bowling alley it is then."

"Catch you there guys," Dave tossed back over his shoulder as we hustled toward base ops to file our flight plan.

Artis's crew had already done the majority of the work for us in base ops. We were able to copy most of the information from their flight plan, which meant we had a few minutes to kill. We elected to walk across the street to the McDonald's, and grab some burgers while we waited for the fuelers to finish pumping on our gas.

We watched Todd, Ray, and Pick land the number three ship just as we were walking out the door of McDonalds. By the time we passed back through base ops to pick up the latest satellite weather photos, Todd's crew had finished shutting down their airplane.

We met Todd's crew halfway across the ramp at almost exactly the same spot where we had spoken with Artist's crew.

Once again, the ramp banter between the two crews was light-hearted and optimistic. Nobody thought that the impending 6.4-hour flight to Lajes would dampen anyone's spirits.

Todd, Ray and Pick gladly agreed to the 0200 rendezvous at the Lajes bowling alley.

A couple of minutes later we were hustling aboard our aircraft, eager to beat Todd's crew off by as much as possible.

All of us were cruising at Flight Level 220, at approximately 290 knots true airspeed. But, Todd's airplane had already proven to be a little stronger than ours.

In C-141s and C-5s, we used to shoot for a specific cruise airspeed. C-141s normally cruise at .74 Mach, and C-5s cruise at .77 Mach. Large turboprop pilots use a little different technique in airplanes like the C-130.

On our old B-model C-130s, the maximum continuous TIT (Turbine Inlet Temperature) at which we can operate the engines is 932 degrees centigrade. That's just about where we run them at cruise. We set about 920 degrees on the TIT gages, and then take whatever airspeed that gives us. It always winds up being something close to 290 knots true.

Some of the airplanes are just a little cleaner, a little less bent, or the engines are not quite as tired, and those airplanes will give you an extra five or six knots of cruise speed. That doesn't sound like much, but over the course of a five-and-a-half hour flight, it makes a difference.

All of our airplanes are equipped with the air-to-air function on the tacan navigation radios. This means that when we're out over the ocean where there are no tacan stations to utilize for navigation, we can select air-to-air and tell how far apart we are from each other.

To make this work the crews have to agree ahead of time who is going to be on which channel, since the tacans have to be tuned to channels exactly 72 channels apart. The standard ploy here is for one airplane to dial in channel 29, and the other to use channel 92. It's easy for pilots to remember, since the standard altimeter setting that we always use above the transition level is 29.92.

We used to use this air-to-air function a lot when we were air-refueling in C-5s. It's a handy way to tell how far you are from the tanker until you're close enough to see him.

On the way to Bermuda, we had constantly watched Todd's airplane gain on us, so we knew he had the fastest ship. Now, we hustled through the checklists to get on our way as quickly as possible.

Soon, I was dropping the condition lever into the RUN position and pressing the start button to light the number 3 engine.

Engine starts in a C-130 are quiet affairs, unlike the C-5 where the pilot often talked his way through a start, reciting exactly what he was expecting to see next.

A C-130 pilot presses the start button, then silently notes whether the light illuminates in the center of the button. If all is going well, depressing the start button will allow bleed air (high pressure air diverted or "bled" from the ground power unit's output) to spin the starter, and the big reliable four-bladed Hamilton Standard props will start slicing through the air.

As soon as the start button is depressed, the crew can hear the ignitors, large sparkplug-like-devices, begin firing.

When the engine RPM reaches 16%, fuel is automatically injected into the jet engine's burner cans, where the ignitors immediately light it off.

This is the same pattern that all jet engines utilize, whether the engine is attached to a gearbox that spins a propeller, or used strictly as a pure jet or turbofan engine. All of them start using the cycle: spin 'em, spark 'em, and spray 'em.

"Spin 'em, spark 'em, and spray 'em" used to be the motto of the T-37 squadron based at Mather AFB in Sacramento, California. It's a handy little memory aid that will keep you on the right track when you're trouble shooting a engine-start problem, though I know that's not what the Mather guys had in mind when they printed it on their patches. But, in view of the ongoing commotion over Tailhook '91, I think I'll let it lie.

Once a jet engine lights off, the remainder of the start sequence consists of checking that oil pressures and hydraulic pressures come on line when they're supposed to, that the engine accelerates properly, and releasing the start button at the proper time. In a C-

130, we release the button when the RPM reaches 60%.

With the number 3 engine up and running and helping out with the bleed air demands, its a simple matter to start the other engine on that wing, number 4, and then spin up number 2 and the number 1 engine.

In short order, we had our before-taxi checklist accomplished, and were commencing the long taxi to the approach end of runway 12 for takeoff.

The airport at Bermuda is a joint-use field, meaning that the civilians and the military share the airport facilities. At Bermuda, the civilian air terminals are located on the southwest side of the runway, close to the approach end of runway 12.

The military facilities are located on the opposite end of the runway and on the opposite side, the north side. It's nearly a two mile taxi from the military air cargo ramp all the way to the takeoff end of runway 12.

Maybe it was just because everyone was in such a good mood—that expectant start-of-the-adventure mood—that I had the flashback. It didn't last long, probably not more than a second. But for a brief moment, I remembered the first time that I taxied a C-5A Galaxy.

I sat next to a clinical psychologist one night a few years ago on a flight from Pittsburgh to San Antonio, and he explained flashbacks to me. He explained it in computer terms.

He said that flashbacks were a way for our brains to store and retrieve valuable data. He said that when somebody says that "Their whole life flashed before their eyes" just before a life-threatening event, that they were really seeing their most valuable memories being rapidly stored in their brain's deepest memory storage locations.

This way, if the brain is damaged, the person will at least be able to remember who they are after the event is passed. It's like a computer copying the data off a floppy disk and putting it onto its hard-disk drive just before an accident.

My whole life didn't flash before my eyes on this day, but for a brief second, I did flash back to 1982. My stickbuddy, Ralph Lucas, and I were going for our first C-5 flight at Altus AFB, Oklahoma.

Our IP (instructor pilot) was a major named Pete. Pete had decided that since we would take our check-ride in the left seat, that we would do all of our flying in the left seat. I was tickled to death with that decision—I couldn't wait to taxi the "aluminum overcast".

On our way out of operations, we bumped into the IP and students who had just finished flying the airplane that we were scheduled to fly. The out-going IP warned us that the nose wheel steering on our airplane was acting up. He said that when the steering wheel was turned, nothing would happen for several seconds, then it would all take effect at once!

We thanked him for the warning, and Pete said that we would go out and take a look at it for ourselves.

Ralph and I tossed a coin to see who would fly first. I won. I could barely control my enthusiasm as I strapped into the left seat of a C-5 for the first time.

Pete took his place in the right seat, and we ran through the before starting engines, starting engines, and before taxi checklists. Ralph observed everything from the jumpseat.

We had two engineers on board, which is the minimum required to fly a C-5. One engineer sits at the flight engineer's station, while the other performs scanner duties.

The C-5 is so large and complex that you have to have a scanner, to run around the airplane and coordinate various procedures. For instance, if you experience a flap malfunction in a C-5, the scanner has to go down the flight deck ladder, make his way back through the cargo compartment, climb up the troop compartment ladder, and then make his way all the way to the forward end of the troop compartment to reach the flap/slat asymmetry control box. It takes a while.

Our two engineers on this day were Ed and Steve. Ed was sitting at the engineer's panel, and Steve was performing scanner duties.

So, with me in the left seat, Pete in the right seat,

Ralph sitting directly behind us in the jumpseat, Ed in the engineer's seat, and Steve playing scanner, I finally pushed the throttles up to taxi a C-5 for the first time in my life.

After a brief stint of straight-ahead taxiing, the marshaller signalled for me to make a left 90 degree turn.

I called for "Caster", meaning that the aft sets of landing gear trucks were allowed to swivel around their struts to cut down the turning radius, and then I started turning the steering wheel to the left.

For a long moment, nothing happened. The marshaller signalled more emphatically for me to start my left turn, and Pete said, "He's ready for you to turn!"

I was just about to answer, when the airplane suddenly lurched to the left. I was too busy straightening out the wheel and placing the aft gear back to the "center" position to answer Pete.

By the time I got everything back under control it was time to make another 90 degree left turn. I called for caster again, and once again turned the steering wheel. Once again, nothing happened.

Now remember that I had never taxied a C-5 before, so for all I knew this lead time between the time that I turned the wheel and the nose wheel actually moved, wasn't exceptionally long.

Pete was just starting to get on my case again about starting the turn when the airplane lurched to the left once more. Again, I was ratholes-and-elbows trying to get the thing going straight again.

By now I knew that this couldn't be right. Pete had made a point of watching my left hand turning the wheel, so he knew that I was doing my part. The airplane was screwing up.

Unfortunately we were now heading for the parallel taxiway at a perpendicular angle, with no way to get back to our parking spot. Our only choices were to stop the airplane and have a tug try to tow us back to parking, which is no small task for a C-5, or try to make the 90 degree turn onto the parallel and taxi ourselves back.

Since the airplane was moving while we were contemplating these facts, our decision was soon made for us. The parallel taxiway drew abreast the nose of the airplane, so I called for caster, knowing there wasn't enough space left in front of the airplane now for a tug to hook onto us.

I started turning the steering wheel. Once more, nothing happened.

The nose of the airplane proceeded across the parallel and out over the grass between the runway and the parallel. Soon, Pete and I had nothing but grass below our seats. The four nosewheel tires are mounted well behind the pilot in a C-5, so I knew they weren't in the grass yet. But, they were getting close.

I was just about to give up and slam on the brakes, when the airplane suddenly lurched sideways into a hard left turn.

I let the ship lurch through more than 90 degrees of turn before centering the aft gear and swinging the nose gear back to the right. If I had timed it just right, I hoped that the nose gear had swung around the taxiway light just to the left of where we had joined the parallel, then swung inside the next taxiway light on its way back to the centerline of the parallel.

When I finally had the airplane centered up on the parallel taxiway, Pete told me to stop the ship. I certainly concurred.

Pete wanted to let Steve (the scanner) disembark to check our tires. If we had run over one of the taxiway-lights, he wanted to know about it before we taxied any longer on cut tires.

I stopped the airplane and set the parking brake. Pete told Steve that he was cleared to exit the airplane to check the tires.

Steve raced down the steep flight deck ladder and flipped the switch to open the crew entry door. Immediately, red hydraulic fluid started spraying past Steve. A short flexible line which carried number-four-system's hydraulic fluid to the crew entry door had split, allowing the 3,000 psi pressure in the system to spray the fluid all the way up to the ceiling.

Steve immediately swung out of the way of the hydraulic fluid to keep the high-pressure stream from cutting him, then he pressed his intercom button and said, "Hydraulic fluid leak at the crew entry door—

shut down the system!"

It was at this point that Ed (the engineer) made one of the classic mistakes in C-5 hydraulic systems.

Each one of the four engines on a C-5 has its own hydraulic pumps which power its own, independent hydraulic system. Having four independent hydraulic systems might sound a little over redundant. But, according to Lockheed, while redundancy is the antithesis of efficiency, it is the refuge of confidence.

Hydraulic systems one and four are the biggest two systems in a C-5. System one contains over 97 gallons of hydraulic fluid!

These systems power so many different components that it's sometimes a little difficult in the heat of battle to remember which hydraulic system powers which component.

One little memory aid that some guys use, is to remember that system four powers the FORward systems, that is components mounted on the forward end of the airplane like the crew entry door.

Unfortunately, another little memory aid that is commonly used is 1-4-1. This refers to which hydraulic system powers each landing gear; system one operates the nose gear and the aft main landing gear, and system four operates the forward main landing gear. Many C-5 crewmembers have also flown the C-141, so this set of numbers is easy for guys to remember.

Ed made the common mistake of remembering 1-

4-1 instead of FOUR-FORward. So, when Steve called him to depressurize *the* hydraulic system for the crew entry door, Ed depressurized system one instead of system four.

Normally this wouldn't have presented a big problem, probably no more than a delay before the proper system was depressurized and a corresponding increase in the amount of hydraulic fluid lost. But, Pete accidentally compounded the problem.

The controls for the brake systems on a C-5 are located on the right side of the cockpit, close to the landing gear lever. The pilot flying in the right seat normally operates these controls.

There are three brake systems on a C-5. The normal system utilizes system four hydraulic system pressure off the forward main landing gear to operate the brakes. The alternate brake system utilizes system one hydraulic system pressure off the aft main landing gear to operate the brakes.

There is also an emergency accumulator which stores hydraulic fluid under pressure, and can be used to stop the airplane with no antiskid protection.

When this whole exercise began we were taxiing with the normal brake system (powered by system four) selected. When Pete heard Steve shouting for the crew entry door system to be depressurized, he naturally assumed that Ed would depressurize system four. So, Pete reached up and moved the brake selector to the alternate position, so that system one would keep the parking brakes set.

Unfortunately, Ed got his four-forward and 1-4-1 memory aids confused, and Ed depressurized system one! This meant that system four fluid continued to spray out of the cracked line by the crew entry door unabated, while system one pressure bled off to zero.

Things started getting pretty confused at this point. Steve was still hanging onto the bottom of the ladder to keep from falling out the crew entry door, and still shouting for Ed to depressurize *the* system.

Ed kept repeating that he had already depressurized *the* system, and that it should bleed itself down in a minute. Pete and I were craning our necks around backwards to try to see Ed's hydraulic panel, to see exactly what was going on there.

It was about this time that Ralph decided to go have a look for himself downstairs. Ralph said "Something must be going on downstairs that we don't know about—I'll be back in a second."

With that, Ralph plugged his headset into the longer scanner's cord, and started down the flight deck ladder. Pete and I were both still trying to get a good look at Ed's panel behind us, paying no attention at all to the taxiway.

The next three things happened so fast that they were almost simultaneous. With my better vantage point from the left seat, I recognized the fact that Ed had the number one hydraulic system depressurized. This was just about the time that Ralph worked his way far enough down the ladder to be able to see out the open crew entry door.

I said, "Isn't that the number *one* system that you have depressurized?"

At almost the same instant, Ralph said, "Why are those taxiway lights whipping by outside?"

Pete and I immediately spun around to look out the windshield, and realized that we were trucking down the taxiway at about 30 knots!

In a sudden panic, Pete immediately reached forward and selected emergency brakes! Since the parking brake was still set, all 24 sets of brakes were immediately slammed on full force!

I can't remember which was louder, the squeal of the tires leaving rubber streaks on the pavement, or the howls of Steve and Ralph as they flew off the ladder and down into the gallons of hydraulic fluid washing about on the floor of the cargo compartment!

We elected to have the airplane towed back to parking.

All these thoughts raced through my mind in an instant as I steered our C-130 down the parallel at Bermuda.

I couldn't stop a smile from spreading across my face behind my boom mike. I considered trying to share the story with the crew, but there was no time. I called for the before takeoff checklist instead.

Ten minutes later we had the gear in the well, racing away from the turquoise-blue waters of Bermuda, toward a date in a bowling alley in the Azores.-

Chapter 4

Six hours and twenty minutes after we put the gear in the well in Bermuda, we were touching down at Lajes Air Base, in the Azores. It was two in the morning, local time.

We had logged a total of 9.7 hours in the air since leaving Charlotte that morning. Everyone was a lot more tired than they wanted to admit.

Our overall detachment commander, Gary Wilfong, had instructed everyone to bring along a small overnight bag for our stay in Lajes. That idea was a life saver. It kept us from having to unload everyone's luggage—all the gear that they were bringing along for a two week stay in Europe.

By the time my crew put our airplane to bed and boarded the crew bus for the ride up the hill to our quarters, we could see Todd, Ray and Pick on short final. In spite of their airplane's faster cruise speed, we had managed to outrun them, although the air-to-air tacans showed that they had steadily gained on us for the entire flight.

My crew hustled up the hill so we could release the crew bus as quickly as possible—knowing that Todd's crew would be ready for it as soon as it made it back to them.

We met Artis, Brendan and Gary in the housing office. All three looked dead tired. Gary had already made certain that everything was okay with all of our rooms, so he was going to bed. Brendan wouldn't let Gary leave before eliciting a commitment from him to meet at noon for a run.

Gary accepted, and, still not having learned our lesson about making these easy commitments, the rest of us chimed in that we would be there also.

Todd's crew walked through the door just as my crew was finishing our registrations.

I don't think anyone was really too hot on the idea of going to the bowling alley at this point, but nobody wanted to be the first to say so. After all, it was the first night of our big adventure.

So, when Todd said, "See you guys in the bowling alley in ten minutes?" we felt to compelled to respond, "We'll be there."

JT was planning a heavy workout the next day, so as we tramped across the dark base to our quarters, JT let it be known that he wouldn't be joining us in the bowling alley. Terry Heath, our nav, voiced similar sentiments.

Dave and I tossed our bags in our rooms and headed for the first rendezvous of the adventure. A spark of life seeped back into our steps as we paced back across the base—anticipating the free-for-all to come.

Soon we were sitting around the longest table in the bowling alley, sipping our first round of cool malted beverages, and trading stories.

Todd mentioned something about how much fuel

he had when he landed, and that reminded me of a story that a friend of mine, Larry Struck, sent to me.

Larry told me that he was flying with a captain who wanted to put 5,000 more pounds of fuel on his DC-8 for a flight across the Atlantic, but a check-captain that was flying along with them didn't want the extra fuel pumped onboard. The captain and check-captain argued back and forth about it for a while, until the check-captain finally said, "Look, you're going to burn 1,300 pounds of that fuel just tankering it across the pond."

Larry's captain said, "You're right! I forgot about that," then he turned to Larry and said, "Call dispatch and tell them to add 6,300 pounds of fuel for me, will you?"

That got a good laugh out of the guys, mostly because we're all so nervous about flying our B-model C-130s across large expanses of water with our limited fuel supplies.

The next time you glance at a C-130, take a moment to notice the teardrop-shaped external fuel tank that sticks down below the wing—usually between the engines. Believe it or not, that external tank holds more fuel than any other fuel tank on the airplane.

Since none of our B-models have those external tanks, our range is much less than most C-130s—hence our nervousness about fuel.

Todd said, "I know our navs are good, but it's

always a good feeling to hear that radar controller on the island tell you that he has you on the scope."

That reminded Brendan of a story that he had heard Rod Machado tell at Oshkosh. Supposedly the day after the Air Force admitted publicly that it did have a squadron of stealth fighters, was the first time that an F-117 turned on its transponder.

You could hear the glee in the departure controller's voice as he called the F-117 pilot and said, "Good morning Bandit 39 — *radar contact!*"

Brendan and I swapped stories for awhile, me telling Larry Struck tales and Brendan relating Rod Machado yarns.

One of my favorite Struck stories is the one about the JAL DC-8 that landed short at San Francisco back in the early '70s. Supposedly the crew knew that the captain was going to land short, but they couldn't say anything to him because he would "lose face".

In fact, after the airplane finally settled onto the shallow bottom of the bay out amongst the approach lights, the copilot turned to his captain and said, "Ahh, look Captain-san—whole damned airport *flooded!*"

That one reminded Brendan of a story that Bill Mulcaha told him about flying into Los Angeles back when Bill worked for PSA. The controller kept calling out JAL traffic to Bill, but he couldn't see him. After the third or fourth call, Bill's captain finally transmitted, "We can't see the Jap!"

There was a long period of silence on the radio, which was finally broken by a distinctly oriental voice saying, "Look into the sun, Yankee dog."

Since Brendan and I both fly the F-100 in Charlotte, we fly with a lot of the same captains. We have a large number of ex-PSA captains flying the F-100, and most of them have great stories about the old days at PSA.

I asked Brendan if he had flown with David Butterfield. He had. I asked him if David had related the normal career progression at PSA to him. He had not, so I did.

David said it was not unusual for a PSA pilot to make captain, get divorced, and buy a big boat. After living on the boat for a year or so, the guys would usually remarry—invariably to a flight attendant.

Hence David's line, "PSA was a great company—when you checked out as captain, they almost always named a flight attendant after you."

Pick thought that one was pretty funny. He also liked Larry Struck's story about the airline captain who finds himself standing in his chief pilot's office, trying to explain his way out of a minor transgression.

After being on the hot seat for the better part of an hour, the captain is just about ready to leave when the chief pilot says, "One last thing—I hear you've been telling everybody that I'm a jerk, is that true?"

The captain says, "It wasn't me chief—I don't know how they found out."

The table gave me a decent laugh for that punch line, but they really howled when Pick interjected the opinion, "That fella shot himself in the foot so many times he was out of ammunition."

Maybe you had to be there—but that observation really brought the house down. Coming at the end of a long satisfying day, I think everyone sensed that we weren't going to top the guffaw that Pick had just given us.

So, we drained our glasses and headed for the door.

It hadn't seemed like we had been in there for very long, but it must have been longer than we thought. The sun was just inching above the horizon as we marched back to our rooms.

Brendan left us with a quick reminder to set our alarm clocks for the noon run, then we split up and made our way to our individual rooms.

If the world had ended in a thundering fiery storm anytime within the next six hours—none of us would have heard it.

Chapter 5

High noon in the Azores found six slightly-bleary eyed pilots stretching along the retaining wall outside our quarters; preparing for our run.

Most of us had actually awakened before our alarms went off. After all, it was four in the afternoon at home.

The riots in Los Angeles over the Rodney King case were still going on back in the states. We had listened to the periodic news reports on the HF (High Frequency) radios on the way over the day before, but all of us were curious about the latest events. That topic dominated the stretching session.

Brendan and Gary stood next to each other and pressed against the wall, stretching their calf muscles and tendons so tight that they probably would have sounded like the high-E string on a guitar if you could strum them. Those two are by far the best runners in our squadron, they always do well in official competitions.

Dave and I were doing our best to shove the wall down the mountain just a few feet away from Brendan and Gary. We knew that if we were going to stay with those two for even part of the run, we had better be as limber as possible.

Todd and Ray had paired up a few feet higher on the wall. Those two have been paired up for a long time. They met in college at North Carolina State, in their ROTC classes.

Then, they were in the same pilot training class at Reese AFB (Air Force Base). After undergraduate pilot training at Reese, they went to Little Rock AFB together to learn to fly C-130s.

After Little Rock they were both assigned to the same C-130 squadron at Pope AFB, North Carolina. After Pope, they both pulled two-year tours at Andrews AFB together.

When their Air Force commitments expired, they both landed pilot slots in our squadron in the Air National Guard.

Their career paths finally took slightly different courses when Ray went to work for Delta and Todd for USAir, but for the most part you can count the two of them as one set. Most biological brothers that I know are not as close.

JT passed us on his way to the gym to pump some iron. I'm not sure what Pick was doing, probably sleeping or reading a book. Lieutenant Colonel Pick Freeman spent his early Air Force career as an enlisted PJ (Pararescueman). He's still in great shape, but nowadays when Pick breaks a sweat, he's usually making money at it.

Gary asked what he had missed the previous night in the bowling alley, which got Brendan started on more Rod Machado stories.

As Brendan talked, I led the way down the hill and out the gate. I was the only one of the group who had run at Lajes before—back during my big-MAC days. (Big-MAC refers to C-141s and C-5s.) The other five were content to let me lead the way out, but everyone was keeping track of the route since we already knew we would probably be coming home in smaller groups.

Brendan told Rod's story about the little boy sucking on his thumb as we ambled down the steep hill toward the flight line. While we listened to Brendan, we were all secretly thinking, "We have to run back *up* this mountain?"

Brendan told about the mother who was at her wit's end, trying to think of a way to get her little boy to stop sucking his thumb.

The lady was particularly concerned about her little boy's habit on this day, because she was planning to take her son on a commercial airline flight to see his grandparents.

She tried every form of bribery, persuasion and coercion she could think of, but the little boy would not drop the habit.

Finally, she discreetly pointed to an obese gentleman and said, "Do you see that man over there?"

The little boy nodded.

His mother said, "Well, do you know why his belly is so big?"

The little boy shook his head, "No."

"His belly is big because he sucks his thumb. His belly thinks that whenever he's sucking his thumb, he's eating. So his belly just keeps on getting bigger and bigger to hold all the food that it think's is coming down to it."

The little boy immediately jerked his thumb out of his mouth and said, "I don't want to get a big belly like that."

Of course his mother was delighted, that is until later that day when she was boarding the airline flight with her little boy.

Instead of taking his seat right away, the little boy followed one of the flight attendants to the back of the airplane. When the flight attendant realized that the little boy was following her, she turned around and bent down to talk to him.

She couldn't bend very far, since she was several months along in her pregnancy, but she bent as low as was comfortable and said, "Did you want to tell me something little boy?"

The little boy nodded but didn't speak, so she pressed him with, "What is it?"

He pointed his finger at her stomach and said, "I know what you've been doing!"

We were just passing through the front gate as Brendan finished his story. The going was fairly level, so I thought I would tell one before I started needing all of my breath for running.

Brendan's little-boy story reminded me of an incident that my ol' stickbuddy from Dover and San Antonio, Dave Comstock, told me about.

Dave is a captain for American airlines, out of Dallas. Dave works in their flight academy.

Dave said that one of the other instructors called in sick one day, so Dave had to pick up his students for the day. Unfortunately, Dave couldn't locate the students' training records.

Finally the chief pilot, Fred, decided to call the sick instructor and ask him where the missing files were located.

Fred dialed the number, then pressed a button to place his phone on "speaker phone" so Dave could listen in and participate in the conversation if he needed to.

The phone had barely started to ring when it was quickly answered by a little boy who whispered, "Hello."

Fred cleared his throat and said, "Is this the Marshall residence?"

The little boy whispered back, "Uh huh."

Fred asked, "Is your daddy home?"

The small voice whispered, "Yes."

"Can I speak to him?"

"He can't come to the phone right now."

Fred arched his eyebrows, glanced at Dave, then turned back to the speaker phone and said, "Is your mommy home?"

Still is a low whisper, the tiny voice said, "Uh huh."

"Can I speak to her please?"

After a brief pause... "She can't come to the phone right now."

Dave was suppressing a smile as Fred tried another tack, "Do you have any older brothers or sisters?"

Still in a whisper, "Uh huh."

"Are any of them home?"

"They're all home."

"Can I speak with one of them please?"

After a painfully long pause, "They can't come to the phone right now."

Fred finally asked, "What are they all doing?"

The little voice was whispering so low that they could barely hear him as he answered, "They're looking for me."

We quickly passed through the small village area just outside the front gate and into the surrounding countryside. Gary and Brendan were setting a pace that I knew would be too fast for me to carry for long.

Dave started singing some crazy country and western song that I had never heard before. The chorus went something like, "I'm so miserable without you it's almost like having you here."

Gary fell back a little so he could run beside Dave. When Dave stopped singing to catch up on his breathing, Gary said, "Hey Dave, do you know what you get if you play a country song backwards?"

Between gasps Dave said, "What?"

Gary said, "You get your girl back, your dog back,

your job back, your pickup truck back..."

That was enough to do me in—I didn't have the breath to keep up their pace and laugh too. We were just approaching an intersection with a road that broke off to the right.

Ray and Todd had opted for a right detour back in the village. I thought that I might be able to link up with them by veering off to the right at this intersection.

Brendan and Gary had already announced their intentions of running all the way to the coast, so I bade them farewell and broke right onto the deserted road.

The sky was clear blue, the terraced fields shimmered in their green abundance behind their ancient stone fences. As I ran alone, breathing in the clean salt-laden air, I couldn't help but feel exhilarated—so alive. So full of anticipation for the adventures I knew were coming, while at the same time so full of the moment. Colors were extra bright. I could feel the different textures of the road and its rich dirt shoulder beneath my shoes.

I could smell the cattle in the fields and taste the salt in the wind, and hear the breeze scuffling its feet in the leaves and grasses blowing over the road around me.

There was no other thought in my mind, other than this, "Thank you God, for this wonderful day."

Chapter 6

I ran for about forty-five minutes without encountering Todd and Ray, or anyone else for that matter. I had been circling back toward the base. I'd just decided to quit running as soon as I reached the next road intersection, when I noticed somebody else approaching the intersection from the right. It was Dave.

He had kept up the punishing pace that Brendan and Gary take for granted until he finally realized that they really did intend to run all the way to the coast.

Dave had also decided to stop as soon as he reached the intersection, but instead we fell in side by side and kept up the pace toward the front gate. To keep our minds off the strain, I asked Dave if Brendan had told any more Machado stories.

Dave said that he had told the one about the pilot that liked to call Atlanta center whenever things were a little slow and say, "Hey center, I've got a PIREP (PIlot REPort) for you."

When center told him to go ahead with the PIREP, this guy would key his mike and say, "It's clear!"

He also told one about a blonde flight attendant who showed up at the hospital with burns on both ears. When the doctor asked her how it happened she said, "I was ironing when the phone rang."

When the doctor asked, "What happened to the other ear?" she tossed her head and replied, "Well, after I burned my right ear, I had to call an ambulance!"

Blonde jokes were all the rage at that time. My blonde daughter, Karen, was always coming home with a new blonde joke back then.

As we passed through the front gate and started running past the flight line toward the bottom of the *big* hill, I told Dave the latest blonde jokes that Karen had told me.

The first was,"What do you call four blondes in a freezer? Frosted flakes."

The second was, "What do you call a blonde in a leather jacket? Rebel without a clue."

The last one was the blonde who claimed that she knew all of the states and their capitals. When the teacher asked her the capital of Vermont, the blonde shook her head from shoulder to shoulder for a minute, then said, "V?"

We reached the bottom of the hill and quickly decided to take the sidewalk to the top. It was steeper than running down the road the way we had come, but it was shorter.

The sidewalk route passed through a small playground complete with swings and teeter-totters. It was the only part of the route that was level, every other step along the route was uphill.

We made it about halfway up the stairs on the uphill side of the playground and finally gave up. It was just too steep. My calves were shaking as we walked up the rest of the sidewalk route and back to our starting point.

Todd and Ray were waiting for us. They had displayed enough common sense not to try to run up the hill. They didn't look nearly as beaten up as Dave and me.

Todd was springing for Cokes from the machine just inside our building, so the four of us sat on the wall for about an hour, drinking Cokes, and waiting for Brendan and Gary to return.

Pick, JT and Artis made appearances while we waited. Everyone seemed to be agreed that the group should take taxis downtown for dinner.

When Brendan and Gary finally made it back, they readily agreed.

Everyone showered and shaved and got into some jeans, and soon the entire entourage was shuffling around the village square downtown, looking for a good place to quaff a beer.

We found an excellent bar with patio seating right beside the ocean. The conversation was as light and airy as the bar.

Something that Artis said about the taxis that we rode in, reminded Dave of an airplane taxi story that he had heard. Everytime I hear this story there are different airlines involved, so I won't even try to use the same names that Dave used. I'll just tell it using a commuter pilot and a major air carrier pilot.

The story goes, that a small propeller-driven

commuter airplane is taxiing out behind a large commercial jet one day. The commuter pilot calls the jet on the ground control frequency and says, "767 on taxiway bravo, this is the commuter airplane behind you, is there a discreet frequency that I can talk to you on?"

The 767 pilots ignore the call, so the commuter tries again, "Hey 767 on taxiway bravo, is there a company frequency or something that we can use to talk to you?"

Again, the 767 pilots ignore him. Finally the commuter pilot calls ground control and says, "Hey ground, can you have the airplane in front of us call us on a discreet frequency?"

Before ground control can answer, a measured, somewhat arrogant voice responds over the radio, "Ground, you can tell that commuter flight that anything they have to say to us can be said on the ground control frequency."

After a long pause the commuter pilot answered, "Okay buddy, your gear pins are still in!"

We walked about a block to a restaurant that the barmaid recommended, where we sat down to a delicious seafood feast.

While we were waiting for the food to arrive, Gary told us about a story that he had heard about a captain named Eddie who liked to taxi airplanes pretty fast."

Eddie was giving his flight engineer a lot of grief on the trip, always playing some sort of practical joke on him. When they picked up a 727 with no APU (Auxiliary Power Unit), the engineer finally saw a chance to pay Eddie back.

When an airplane does not have an operational APU, standard procedure is to start one of the engines using an external air supply to spin the engine, then use the operating engine to supply air to crank the other motors.

The external air supply is usually an air bottle, a huge cylinder on wheels that looks a lot like a WWII atomic bomb. The compressed air contained in the cylinder is usually just enough to crank up one jet engine before it has to be refilled.

When the airplane was all closed up and ready to go, the copilot received permission from ground control for the ground handlers to push them away from the gate and out onto the ramp, where the air bottle was plugged in for the start.

They were running a little late, so as soon as the first engine was on speed, Eddie waved off the ground handlers, released the parking brake, and started taxiing across the ramp. He intended to start the other two engines on the taxiway, enroute to the runway.

The ground handlers disconnected the air bottle, just like they were supposed to, but the flight engineer knew that Eddie had never heard anything over the intercom that told him that.

Eddie was taxiing a little fast, as usual, when he

said over the intercom, "Okay, let's get those other two engines cranked up."

There were no ground handlers still on the intercom of course, but if they had still been plugged in with their external cords, they would have heard Eddie's statement.

The flight engineer turned the variable air nozzle on his panel so that it would blow directly onto his microphone—making it sound very similar to what a microphone would sound like outside, close to the running engine.

Then he cupped his hands around the microphone to further the effect, pressed his floor mike switch and said, "Hey captain, are you ready for me to disconnect this air bottle yet?"

Thinking he was racing across the ramp towing the air bottle cart behind him by the air hose, Eddie immediately slammed on the brakes!

It didn't take him long to figure out that he had been had, but that didn't help him explain it to the flight attendants who had nearly been thrown to the aisle by his sudden stop. Eddie didn't get any more coffee from the flight attendants for the remainder of the trip.

We savored the seafood, listened to the stories, and generally enjoyed a great evening of camaraderie. As soon as everyone had finished we paid the bill and corralled some taxis for the ride back to the base.

The night was still semi-young, but we knew we

had a full day ahead of us tomorrow.

We planned an early departure for the flight to Belgium.

Chapter 7

We were all awake before sunrise Sunday morning. My crew was scheduled to make an 0820 departure. Everyone was anxious to be on our way. I knew we would be leaning forward in the straps long before our scheduled takeoff time arrived.

After a quick shower and shave, I grabbed my overnight bag and met the rest of the entourage outside the building. Our rooms were equipped with small coffee makers, so everyone had a steaming cup of coffee in their hands as we sat on the retaining wall, waiting for the crew bus to pick us up.

We got to watch the sun come up while we waited. It was beautiful.

After checking out of the quarters and paying our bill, we sortied down to the airplane to drop off our bags and ensure everything was progressing normally there, then we hopped a ride back up the hill to the Lajes mess hall.

Artis was one of the first in our group to reach the grill, and he ordered something called an aviator omelette. Basically, an aviator omlette contains one part of everything that the cook can lay his hands on.

The approved accompaniment for the omellete was hash browns and toast, with plenty of hot sauce spread over the entire plate. Don't forget the extra cup of coffee to keep that hot sauce going down, and you've got yourself a breakfast guaranteed to see you through the 5.4-hour flight to Belgium.

After breakfast, all three crews flight planned together down the hill in base operations. This was the same base operations that I wrote about in my first book—the place where I hitched a ride with a startled McGuire crew after my Travis crew had to leave me behind. The place held a lot of memories for me.

The plan was for each of us to arrive in Belgium with 10 minute spacing. We made certain that all three crews were flying the same route, same altitude, with appropriate entry times to the oceanic flight information regions.

I won't go into great detail as to the various hijinks that accompanied the three-crew flight planning session, but suffice it to say that with everyone in high spirits and well rested, the flight planning activities were anything but droll. I wish I had a video recording of that morning.

Once again, Artis, Brendan and Gary led the charge, lifting off exactly on schedule at 0810 local time. We were next.

After a harried bit of last-minute chart thievery from Todd's crew, we were ready to turn engines on schedule. I didn't feel too badly about Terry stealing the charts from Todd's navigator, Bob Kotula. After all, Bob had stolen them from base operations.

When tower issued our takeoff clearance JT pushed up the throttles. Those big Hamilton Standard props bit into the dense sea-level air, drawing us forward ever faster until JT eased back on the yoke, causing our straight-edged high lift wings to flex upward, take our weight on its spar, then rise up sweetly into the clear blue Atlantic sky.

During our takeoff roll I glanced over at the ramp, and saw Todd's entire crew standing in front of his plane, waving their arms and pumping their fists—urging us on with the enthusiasm of teammates who have just witnessed an interception.

I knew how we looked to them—the same way that Artis's airplane had looked to us—great!

We cleaned up the airplane on schedule, gear up—wait until they're up and locked to keep from overtaxing the utility hydraulic system—then flaps up. Retract the landing lights back inside the wings, then accelerate to climb speed, 180 knots up to 10,000 feet.

A half hour later we were cruising at flight level 210. As often happens when flying over oceans, the radios grew quiet for long periods at a time. We only needed one of our two HF (high frequency) radios to monitor Shanwick oceanic control, so I decided to tune the other to the BBC's HF frequency to catch up on the latest news.

As I leaned forward and dialed in 150700 in the frequency selector window, my mind flashed back to another day about fifteen years ago, when selecting a radio frequency in an Army helicopter I was flying had very different results than the one I was hoping for.

It was probably the spring of 1978, though I can't recall with certainty exactly when it happened. I do recall that I was flying an NOE (Nap Of the Earth) mission with my good friend from flight school, Hal Johnson.

Both of us were young Army warrant officers, in our early 20s. Our mission was to flight-check an operations area in Fort Hunter-Liggett for NOE flight. Our unit was scheduled to participate in a camouflage-detection test in the area, which would require us to fly our helicopters at NOE altitudes (read skids-in-the-weeds).

Before we operated at NOE altitudes, we always performed a flight check of the area. This normally involved one scout helicopter, an OH-58, flying at NOE altitudes in a grid pattern over the prescribed area, while a second OH-58 flew cover about a thousand feet overhead. The cover ship was responsible for helping the low ship spot hazards, and being on site and ready to call crash rescue and vector them exactly to the low ship, in case the NOE bird didn't spot a hazard in time.

Our mission started that morning at Fritzsche Army Air Field, at Ft. Ord. Hal was assigned one OH-58, and I another. The plan was for us to fly down to Ft. Hunter-Liggett in formation, perform the NOE flight check, and then drop my helicopter off at the airfield at Ft. Hunter-Liggett. I would catch a ride back to Ft. Ord with Hal.

The first little wrinkle in our plans occurred when one of the operations clerks ran out to Hal's helicopter just after he finished starting his engine. He handed Hal an envelope that he wanted dropped off at Hunter-Liggett.

Unfortunately, this interruption occured at precisely the moment that Hal was twisting his throttle to the full open position.

The interruption caused Hal not to notice that the engine acclearated all the way to 6,600 rpm, instead of stopping at 6,000 rpm like it was supposed to.

The fact that the engine accelerated all the way to 6,600 rpm indicated that the linear actuator on the throttle had failed, but with the operations clerk yelling in Hal's ear, he didn't notice it.

I supposed if you've never heard the term throttle linear actuator before, that might sound a little intimidating, but it's not really as complicated as it sounds.

You see, the turbine engine in our OH-58 helicopters operated at 6,600 rpm. The engine was started with the throttle rolled back to the flight idle position. When the pilot twisted the throttle counter-clockwise with his left hand, the engine speed increased, up to a maximum of 6,000 rpm.

The engineers who designed the helicopter didn't trust the pilots to crank on the last 600 rpm with the throttle. They felt like it would be too easy to twist the throttle a little too far and accidentally overspeed the engine, as well as the transmission and rotor system.

So, the engineers mounted a little gray box at the end of the throttle shaft called a linear actuator. Inside the box was a small electric motor with a gear on the end of it's shaft. That gear meshed with a toothed rod which finally connected with the fuel control.

The pilot had a little toggle switch mounted on his collective lever, just above the throttle. When the toggle switch was toggled back (beeped was our term), the electric motor retracted the toothed rod back into the linear actuator. In this position, the fuel control would only allow the engine to run at 6,000 rpm.

After the engine was stabilized at 6,000 rpm, the pilot would beep the toggle switch forward. The little electric motor in the linear actuator would extend the rod very slowly, until the engine was finally stabilized at 6,600 rpm.

Once the linear actuator was set, the pilot could roll the throttle back to idle, and then twist it back to full open. With the linear actuator already set, the engine would accelerate all the way back to maximum continuous speed, 6,600 rpm.

Before the pilot started the engine, the checklist called for him to hold the linear acuator toggle switch in the retract position for 10 seconds, to ensure that the rod was fully retracted. Hal had done that, but unbeknownst to him, the little gear on the end of the electric motor in his linear actuator had sheared off—leaving the toothed rod free to retract or extend as it pleased. It just so happened that it stayed in its last position, which was set to produce full engine speed, 6,600 rpm.

After the operations clerk cleared his rotor system, Hal resumed his checklist. Since the engine was already stabilized at 6,600 rpm, Hal assumed that he had already beeped the linear actuator up to full speed.

We made a formation takeoff and flew down the Salinas valley together at 1,000 feet AGL (Above Ground Level). Normally we didn't fly that high, but when transiting the valley we would pick up the altitude a little bit to cut down on noise complaints.

After about a half hour we made a right turn into the draw that led to the northwest pass—the airborne entry point into Ft. Hunter-Liggett.

We had to climb steeply to gain enough altitude to make it through the pass. After clearing the saddle at the top of the pass by less than a hundred feet, we began a steep descent into Hunter-Liggett.

We flew past the airfield, continuing south to the operations area we were tasked to survey. Since I was leading, I opted to go low first. So, Hal picked up his altitude to about a thousand feet above me, while I began a systematic NOE check of the area.

There's nothing in the world quite like flying a helicopter at NOE altitudes. When I fly a big airplane, I have no desire to cowboy the airframe around. For me, it would be like racing a Cadillac over a motocross course. The machine just wasn't built for that.

But, let me strap a helicopter to my butt, and all at once spurs start growing out of my heels. I feel an almost irressestible urge to pull some G's and yell, "Yeeehaa!"

I believe there are certain glands and organs in a pilot's body that appear to medical science to have no useful purpose—such as the appendix. But, put that pilot in a helicopter with his skids in the weeds, and those glands start squirting out fluids that make testosterone seem like mother's milk. It's a rush.

I didn't want to hog all the fun, so after about half the area was finished I climbed up to takeover the cover ship duties and let Hal have some fun.

Hal Johnson is a great NOE pilot. Hal surveyed the rest of the area in quick order, completely unaware that his low level maneuvering was running the risk of shaking loose the free floating toothed rod in his linear actuator.

Whenever that rod broke lose, the fuel control would be free to either overspeed, or fall back to the normal full throttle position of 6,000 rpm. If that happened while Hal was brushing his skids through an oak tree in a 45 degree bank, he was a gonner.

As soon as Hal finished his survey he climbed up on my wing and we flew back to the airfield in formation. After a quick circling approach, I hovered to the prescribed parking spot and started shutting down my bird.

Hal set his ship down on the engine-running pad and called for someone from operations to run out and pick up the envelope he had been given at Ft. Ord.

By the time the operations clerk picked up the envelope from Hal, I had finished tying down my rotor blades. I grabbed my helmet bag and ran to Hal's 58.

I clambered into the left seat and strapped in, and in short order Hal was chauffering me back through the northwest pass and down into the Salinas valley.

We always had to watch out for hawks and eagles as we transitted the northwest pass, they would often dive on us. It was as if they perceived us as competing birds of prey, and wanted to drive us off.

Soon, we were cruising along at 90 knots, about 1,000 feet above the Salinas river. I glanced at the clock on the instrument panel and realized that it was almost 1130. That was the time that the "Gong Show" used to come on in California.

The FM radios in our helicopters were capable of receiving TV signals. We normally used frequencies in the lower band of the radio's ability for our communications, such as 30.30 ("Winchester"). In order to listen to TV stations, we had to dial the radio up to the upper limits of its frequency capability.

The Gong Show came in on frequency 73.75 on

our FM radio.

Of course we couldn't see any picture, but a lot of times people would say funny things on the show, so we liked to listen to it.

I reached down and started dialing in 73.75 on the FM radio. The show always started with a raucous brass band playing a very loud, attention-getting tune.

As soon as I finished dialing 73.75 in the radio's frequency window, I immediately heard the loud brass band playing the show's opening theme song.

Much to my surprise, Hal immediately shoved the collective to its full down position and rolled the throttle back to flight idle. He had to lower the nose to keep the airspeed up, because we were now in a full-fledged autorotation!

We had been flying over the river, since it was the least noise sensitive area. The river's banks were covered with trees on both sides. Since we were only starting from a thousand feet up, we didn't have enough altitude to be able to glide to a clear landing site.

There was a narrow sand bar in the middle of the river in front of us. Hal's only chance of completing the autorotation without putting the 58 in the water was to try for the sandbar.

Hal was busy transmitting mayday calls on 243.0, the emergency frequency. As he quickly ran through the engine failure checklist, I realized that he was rapidly approaching the step which required him to depress the throttle's flight idle button, and roll the throttle to the full off position, in order to lessen the chances of a post-crash fire.

While Hal was busy making the mayday calls, I was quickly scanning the engine instruments. The engine was running, although it was now running at idle since Hal had rolled the throttle back to flight idle.

I quickly realized what had happened. I hadn't told Hal that I was tuning in the Gong Show. When the brass band suddenly sounded in his headset, he must have thought that it sounded just like the low rpm audio warning—the warning that sounds if the engine quits and allows the rotor rpm to deterioate.

I smiled, thinking, "This is going to be great! Hal thinks the engine has quit and he's about to ditch this helicopter, when actually all that is wrong is the theme song of the Gong Show is sounding in his headset!"

Hal was rapidly running out of airspeed, altitude, and ideas. We obviously weren't going to make the sandbar—he was going to have to complete the autorotation into the river. Nobody was answering his mayday calls, and instead of helping him, I was hanging onto the throttle with a death grip to prevent him from rolling it completely off and actually giving us a real engine failure.

I waited until Hal finished another mayday call,

then I pressed my intercom button and calmly said, "Hal."

Hal tore his eyes away from the river and looked and me as he answered, "Yea?"

I looked him straight in the eye and said, "Hal— *it's the Gong show!*"

I'll never forget the look on Hal's face. He didn't even have his FM button pulled up on his intercom box, so he couldn't even hear the brass band playing. He actually *had* heard the low rpm audio sound, and entered autorotation exactly as he was supposed to.

What he hadn't realized, was that the low rpm audio was sounding because the toothed rod in the linear actuator had finally broken lose, and allowed the fuel control to roll the engine back to 6,000 rpm. The low rpm audio sounded at 6,100 rpm, well below the engine speed required for normal flight.

But, all of this linear actuator information was well beyond Hal's present scope of operations. As far as Hal was concerned, he had experienced an engine failure over unrecoverable terrain, been unable to contact anyone on the emergency frequency, and his old pal Sherm had completely gone bonkers on him at the first little sign of trouble.

After allowing himself the briefest moment of distraction with me, Hal quickly went back to work flying the helicopter.

He was still trying to roll the throttle completely off, as the checklist called for.

I broke into his realm of attention again by saying, "Hal, the engine is still running."

Hal quickly responded, "No it's not, the needles are split."

The engine and main rotor tachometers were set up so that when everything was on speed, their two needles were aligned, one on top of the other, so that they looked like one needle on the tachometer gage. With the engine now operating at idle, it's rpm needle was separated from the rotor's needle, hence the term "split needles".

I said, "They're split because the throttle is rolled back to flight idle, but look at the other engine instruments—the engine is running at idle!"

That confused Hal. First I had given him some totally bogus information about the Gong Show, and now I was pointing out that the engine hadn't quit after all.

I was starting to get concerned. We were getting awfully close to taking a swim.

Hal said, "I heard the low rpm audio."

I said, "Just roll the throttle up! The engine is running, roll the throttle up!"

He did. I relaxed my grip and Hal quickly twisted the throttle back to its full throttle position. The engine quickly accelerated to 6,000 rpm.

Now I was confused. I didn't understand why the engine hadn't accelerated all the way back to normal rpm, but we didn't have time to work it out.

Hal started raising the collective lever to increase the pitch in the main rotors and stretch our glide to the sandbar. The main rotor immediately started slowing down, until it's tachometer needle met the engine's needle. Now the engine took over and prevented the rotor rpm from deteriorating further.

But, the rotor was turning well below its normal flight rpm. The blades, which depend on the centrigual force of their rotation to maintain their rigidity, were coning up so far above the helicopter that I was afraid that they would break off.

Hal skillfully used just enough power to stretch the glide to the sandbar, then quickly eased the 58 onto the ground in a sort of semi-powered autorotation.

As soon as we were safely down, Hal frictioned down the control locks on the collective and cyclic, then turned to me and said, "What the hell does the Gong Show have to do with any of this?"

We eventually figured out what had happened and how we could recover the aircraft without being found out. We swore each other to secrecy for as long as we both remained in the Army. For the next several years, the "Gong Show" was our private little inside joke.

It made me smile to remember it again, as I dialed 150700 into the HF radio.

Soon the sounds of the BBC were beaming into

my David Clark headset. I retrieved my cup of coffee from its metal holder, reclined my seat and settled in for the 5.4-hour flight above the sparkling Atlantic.

Chapter 8

The white cliffs of Dover disappeared under our left wing as we banked to the right to take up our new course for Koksijde Air Base. Soon the dark sands of the beach at Dunkirk were visible.

Terry, our nav, adjusted our heading slightly to the north, and we quickly picked up the runways of the Belgian Air Force air base.

The air traffic controllers had held us rather high, so we made a steep circling approach to lose altitude and align ourselves for landing. A few quick chirps from the main tires, and we were chasing the "follow me" truck to our parking spot.

There isn't a lot to the air base at Koksijde. Our operations center was set up in a WWII-vintage hooch, which apparently spends the rest of the year vacant. **

After unloading all of our professional gear and storing it in the operations hooch, we were shuffled onto buses with our B-4 bags and chauffeured to the other side of the field for our in-country briefing.

After a couple of hours of warnings about all of the possible mis-steps that awaited us, we were dismissed to proceed to the Belgian dining hall for lunch.

Most of us were harboring secret wishes for a plate full of authentic Belgian waffles for lunch, but we were sorely disappointed. A few of the guys even opted for the MREs (Meals Ready to Eat) that were being passed out in the back of the mess hall.

Of all the complaints I heard in the chow line, Dave came up with the best one-liner. When we reached the "meat" table, a Belgian enlisted man started to fork a huge slice of canned meat onto Dave's plate.

Dave took one look at the assortment of congealed fat, salt and meat cubes, and quickly drew back his plate, exclaiming, "What is that?!"

The Belgian replied, "Calve's tongue."

Dave wrinkled up his nose and said, "I'm not eating anything that comes out of an animal's mouth."

The Belgian shrugged and asked, "What would you like then?"

Dave said, "How about some eggs."

We quickly finished our lunch and joined the que waiting for bus rides to our new home—the Holiday Park in downtown Koksijde. The Holiday Park is a resort community located just a block from the sea shore. We were assigned two-bedroom suites in the Barkas and Bravos buildings.

Each suite was set up to house three crewmembers, one in the master bedroom, and two more doubled up in the second bedroom. Terry was a major, while Dave and I were both captains, so we let Terry have the master bedroom.

Our suite had a combination kitchen/living area on the first floor. The second floor contained the master bedroom, and the only bathroom. Dave and I were doubled up in the tiny space upstairs which functioned as the second bedroom.

Actually, the space had been recovered from the attic. A ship's ladder led up to the tiny space, where two cots were sandwiched up under the steeply sloping leaves of the roof. Even though our cots were sitting flat on the floor, it was impossible to sit up straight on them without hitting our heads on the steeply-sloped ceiling. We later learned that Dave and I shared the smallest bedroom in the entire resort.

The rest of our crew was bunked out two doors down from us. Being the senior enlisted man, Jim Young had immediately requisitioned the master bedroom for himself, leaving Ned Seaman and Tommy Parsons to double up in the second bedroom.

In their suite, the second bedroom was arranged under the master bedroom, which resulted in an embarrassing incident for Tommy during their first night in the resort.

Tommy was sound asleep in the upper bunk of their bunk bed, with Ned sleeping lightly in the bunk below. All at once, Jim rolled onto his back in the bedroom above, and commenced one of his world-famous rounds of power snoring.

Tommy instantly sprang out of the bed, groping for his survival knife and letting out his best rebel yell. Ned rolled out of his bunk and swiped his hand across the light switch, while brandishing a boot against the unseen intruder.

After a few seconds of eye-blinking muscle-tensing hysteria, Tommy and Ned realized that they were faced off against each other. Ned was just about to ask Tommy what was going on, when Jim expelled another window-rattling snore over their heads.

An intuitive smile quickly spread across Ned's face as he watched Tommy's eyes light up with understanding.

Tommy looked up at the ceiling, then back at Ned and said, "Sorry buddy—I thought they were cutting through the roof with a chainsaw to get to us."

Ned just laughed and tossed a package of ear plugs to Tommy so they could go back to sleep.

**(Actually I learned later that the "hooch" I described at the beginning of this chapter was the old Belgian Air Force 40th squadron's operations building—constructed just after the German night fighters vacated Koksijde near the end of WWII. A thousand missions of mercy were launched out of this old building, and flown by brave Belgian pilots willing to risk their lives to rescue sailors in peril in the treacherous North Atlantic. Their new operations building is still a hubbub of activity.)

Chapter 9

Monday morning, the 4th of May 1992, dawned clear and bright on the coast of Belgium. Sunlight streamed in through the small window sliced into our attic bedroom, causing Dave and I to stir at first light.

We were scheduled to attend two briefings at the airbase, one at 10 o'clock and the other at 2 in the afternoon. We knew we were waking up way too early.

I rolled out of my cot first, being careful not to rise too quickly lest I crack my noggin on the low ceiling. I negotiated the steep stairs with a minimum amount of noise, and proceeded to the kitchen to start boiling water for coffee and tea.

By the time I had the kettle on the burner I could hear Dave in the bathroom overhead. I looked out the window at the clear blue skies, and realized immediately that today was going to be an adventure—no matter what.

Terry beat Dave downstairs by a couple of minutes. Terry and I replayed our tennis game of the prior evening until Dave showed up, when the conversation switched to the quality of sleep in Belgium.

Dave started it by asking, "Did you hear those cats last night?"

I shook my head, "No," but Terry answered, "Yea! Just after midnight—I thought they were going to tear the place down!"

Dave nodded and said, "I've never heard cats that loud before. I thought it was a mating call for a car alarm at first."

The tea kettle started to whistle, so I busied myself pouring water over the instant coffee and tea bags arranged in the cups on the counter.

Terry took the opportunity to tell the story about the salesman who knocks on a farmer's door to tell him that his cat has been run over in the road.

The farmer says, "How do you know it's my cat?"

The salesman says, "You have the only house for five miles around here—it must be your cat."

The farmer says, "Oh yea—what does he look like?"

The salesman screwed his face up into a grimace and drew his curled hands up under his chin in imitation of a deceased feline.

The farmer says, "No! I mean what did he look like before the car hit him?"

Terry shoved both arms out straight and popped both of his eyes as wide open as they would physically go—a perfect imitation of a cat caught in a car's high beams.

Now before you take offense reading this, let me just tell you that all three of us own cats. We like cats—but we also like cat jokes.

By now all three of us were sprawled across the furniture, bare feet on the coffee table, enjoying our coffee.

Terry's joke reminded Dave of the story about the wino who discovers a cat that has been hit on the corner of Sycamore and 42nd streets.

Being an animal lover, the wino drops a quarter in the pay phone on the corner and calls the humane society, in hopes of rescuing the injured kitty.

When the humane society worker answers the phone, the wino says, "I'd like to report a run-over kitty."

Unfortunately, the gentleman is slurring his words a bit—nearly to the point of being unintelligible.

The worker replies, "Did you say you want to report an injured kitty?'

"Yea, that's right. A run-over kitty."

"I see. I kitty that has been run over by a car. Very well, what is the injured kitty's location?"

"It's at the corner of 42nd and Sysackamoorea."

"The corner of 42nd and where?"

"Forty-second and Sickisemeora."

"I'm sorry sir—I can't understand you. What is the crossing street at 42nd."

Now with an unmistakable note of exasperation in his voice, the wino says, "Hold on. I'll call you back in ten minutes."

The worried worker pleads, "Where are you going?"

The wino spits into the phone, "I'm going to go kick this damn cat down to Elm street!"

Terry had opened the front door a little to let in

some fresh air, and our laughter at Dave's shenanigans rippled across the little courtyard to Artist's and Brendan's hooch. We were still smiling over our coffee when they appeared in the door, inquiring as to the source of our merriment.

I poured a couple of cups for them while Dave retold his wino/cat tale, drawing a even bigger laugh the second time.

The cat stories reminded me of a yarn that Larry Struck had sent to me some time ago, concerning a friend of his named Jim who owned a small private airplane, a Cessna 150.

Jim also owned a cat of immense proportions. The neighbors suspected that the cat, "Duke", was at least part mountain lion. But, Duke was legendary in the neighborhood for his easy-going personality and friendly disposition, so the neighbors tolerated his presence.

One Saturday Jim was just jumping into his jeep to run to the airport for a little morning spin in his 150, when his neighbor Bill shouted a greeting to him. Jim invited Bill to ride along on the morning jaunt, and soon the two of them were high-tailing it to the airport.

Unbeknownst to them, when Bill entered the passenger's side of the jeep, Duke had also jumped aboard. Duke secreted himself in the back seat of the jeep until Jim killed the engine at the airport, when Duke let out a plaintive meow.

Jim didn't want to leave Duke locked up in the hot jeep, and he didn't want to lose him by letting him run wild around the airport, so he decided to take him along for the ride. After all, what could it hurt?

A top-off at the gas pump and a quick preflight, then Jim, Bill and Duke were on their way to the end of the runway.

Duke sat patiently on the storage shelf behind their seats while Jim pushed in the throttle and urged the little Cessna down the narrow runway and into the sky. Duke seemed to enjoy the climb out and cruise, not even wincing when the power was adjusted or the aircraft's nose was raised or lowered.

After a few minutes of sight seeing, Jim had an idea.

Jim said, "Have you ever heard of the cat and duck theory of instrument flight?

Bill laughed and shook his head, so Jim explained, "The theory goes that if you watch a cat's head, you can always tell which way is right side up. And of course, ducks don't like to fly in the clouds, so if you throw a duck out the window and follow him, he'll lead you to clear skies."

Bill said, "Well, we're fresh out of ducks, but we happen to have a superb specimen of feline grace on the back shelf. Do you want to try it?"

Jim said, "Sure, put old Duke up there on the glare shield and we'll see if the theory works."

Jim held the airplane steady while Bill reached

around to gather up Duke in his arms, then gingerly deposited him on top of the glare shield in front of them.

As usual, Duke took the seat reassignment in stride, settling into his new position without complaint.

Jim grinned and said, "Let's try a left turn."

Gently turning the yoke to the left, Jim caressed the little airplane into a gentle left bank. To their delight, Jim and Bill watched Duke's head remain perfectly vertical while his body bent to conform with the glare shield in the steady turn.

Jim said, "Let's try a climbing right turn."

Reversing the yoke and easing it back, Jim nursed the 150 back to the right and brought up her nose until she was stabilized in a climbing right bank.

Once again, Duke corrected nicely, tilting his head to keep it vertical, and simultaneously tilting his backwards to keep it level.

Bill laughed aloud and said, "That's great! I wonder what he would do in a spin?"

The airspeed was already deteriorating in the climb, so Bill eased the throttle back and said, "Let's find out."

The airspeed needle quickly unwound when the power came off, and soon the stall warning horn buzzed in their ears. Duke took on a slightly perplexed expression.

The critical angle of attack was exceeded on the

left wing first because of the right turn, causing the plane to snap over to the left. The nose rapidly fell through the horizon, pitching the 150 into a full-blown spin.

The nose fell so rapidly that the glare shield was pitched out from under Duke's searching claws. Left to grope nothing but air, Duke emitted an ear-splitting screech and started clawing the air with all four feet as fast as his legs would churn.

Unfortunately for Jim, Duke managed to fall slightly faster inside the cockpit than the airframe was descending outside, which meant that he was going to come down on something.

Much to Jim's chagrin, Duke managed to come down completely astraddle his head!

With all four paws clawing for their lives and his throat screeching for salvation, Duke settled down atop Jim's skull, determined to dig in enough of a claw hold to sustain him through Jim's next gyrations.

At this point, Jim realized that he no-kidding had a problem. Not anticipating the way things had gone, he hadn't started the spin from an exceptionally high altitude—which meant that he didn't have a lot of time to get himself out of it.

His screams had apparently had no effect what-so-ever on Duke thus far, nor had Bill's for that matter. It was obvious that Duke was digging in for the duration, so asking Bill to *pull* him off was obviously not much of an alternative.

There was only one quick fix. Jim shouted, *"Knock* him *off*!"

Now remember that Duke was considerably larger than your normal cat, or your standard cocker spaniel for that matter. With his teeth bared and his claws slashing, Bill didn't really care much for the idea of taking a swipe at the big cat.

But, Bill didn't know how to fly, so he really didn't have much choice. He drew back his left hand, and let fly a back-handed blow designed to remove Duke from his perch.

Unfortunately, the cat saw it coming!

Duke quickly scampered around astraddle Jim's left ear, so that Bill's blow landed squarely on Jim's right ear! To add insult to injury, the big cat got in a slashing blow of his own to Bill's hand before he could draw it back, raking three long red gashes rom Bill's wrist to his knuckles.

With his back claws now curled around Jim's throat and his left front claw gripping Jim's nose, Duke didn't present much of a target for Bill's next shot.

But, it was a blood feud now with Bill, who closed his fist and let fly a jab designed to dislodge the cat's nose hold.

"Crack!" went Jim's nose. "Reeeooww!" went Duke. "Oops!" went Bill.

With his right eye now unimpeded, Jim believed

that he saw a chance for salvation. He quickly raised his left hand to release the left window latch, then stomped as hard as he could on the right rudder to break the spin.

The ensuing lateral forces were enough to break Duke's choke hold, tossing his hindquarters outside the open window. As he yanked back on the yoke to break their dive, Jim reached over and pulled the window closed just enough to prevent Duke from falling the rest of the way out, or from re-entering the cockpit.

The 150 pulled out of its dive just above the tree tops, with Duke, Jim, Bill, and the straining engine all roaring their approval.

Jim never took Duke on another airplane ride. Nor, for that matter, did he invite Bill.

The time crept up on us and we had to break up the bull session to get ready to attend the morning briefing.

The transportation system experienced a melt down sometime during the morning, resulting in about 60 of us crowding onto a 30 passenger bus in order to make it to the airfield for the briefing.

Once again we tolerated the endless warnings about all the bad things that could possibly happen to us, until the briefers finally got to the subject near and dear to our hearts—the flying schedule.

There were two exercises planned for the next two days, but both of those required aircraft equipped with SKE (Station Keeping Equipment), which our aircraft didn't have.

So, we were told that all of the B-model C-130 crews would be released until Thursday evening, when we would be briefed on our missions to be flown over the weekend.

Some of our contingent was ready to hit the road immediately, on the premise that it would be best for us to get out of town before they changed their minds and found something for us to do for the next couple of days. But, by the time we found rides back to the Holiday Park, it was late.

We decided the best course of action would be to enjoy a good dinner and get a good night's rest, and then begin our sojourn fresh in the morning.

To get a little exercise, Brendan and Dave caught the trolley to Oostende, Belgium and ran home. It was about twelve miles.

Terry and I played tennis until our arms were ready to drop off.

The rest of the crews amused themselves as they pleased until it got dark, when everyone drifted over to our hooch for a spaghetti dinner—prepared by "Chef Jim". I'll say this about Jim Young—the man can sure make good spaghetti.

While the guys sprawled around the hooch, seeing who could absorb the most spaghetti sauce in one sitting, the conversation drifted back to our cat stories of the morning.

We retold the good ones for the guys who hadn't heard them that morning, and even managed to include a few new ones that I probably shouldn't repeat here.

When Dave and I finally crawled into our cave in the attic, we were both ready for a good night's sleep. I know I had a smile on my face when I drifted off— anticipating what the morrow would bring.

Chapter 10

We were up with the sun the next morning, Tuesday, May 5th, 1992. The banter flew light and lively as we munched our way through the last of our fruit breakfast, knowing it would be too old to eat by the time we returned.

We were just finishing up the last of our breakfast when the rest of the guys showed up. Everyone except Brendan grabbed a seat while they waited for us to clean up. Brendan left the door open and leaned against the door frame, enjoying the morning air.

The Holiday Park had bicycles for rent, but none of our group took advantage of that opportunity. The bicycles were ancient road warrior models, painted solid black. While Brendan stood guard, he noticed a remarkable sight.

An elderly lady dressed completely in black had rented one of the bicycles, and she was pedaling it past our front door on her way to the beach front road.

Without saying a word, Brendan started humming a tune as the lady passed, "Da da dee, da dee dum, da da dee, da dee dum."

As soon as I looked up and saw the lady, I recognized Brendan's tune as the music from Wizard of Oz. I'll bet we laughed for five minutes.

Finally, we had the dishes cleaned up and the last of our coffee drained from our mugs, and we were on our way.

Terry, who had decided not to join us, called after us, "Where are you guys going?"

We all started laughing at once.

We honestly didn't know.

Terry might have misread our laughter as an insinuation that we knew a secret which we weren't letting him in on. But we truly didn't know how to answer his question.

I shouted back, "We'll call when we get there."

Realizing that was all the answer he was going to get, Terry shrugged and closed the door.

We quick-stepped the short block to the trolley stop, making last minute equipment checks as we went.

"Did you bring any sunscreen?" someone asked.

"Where are we going that we're going to need sunscreen?"

"The Riviera?"

"Yea!"

"I want to go to Moscow."

"Yea!"

"Moscow?! Are you nuts?"

"Hey—if we make it to Moscow, they'll talk about this trip for the rest of our careers."

"Which probably won't be all that long."

"I say England. Let's go to England."

"Yea!"

"Why England?"

"Why not?"

On it went, everyone caught up in the rapture of having time and money and friends, and the world spread at our feet—at least for the next couple of days.

The trolley was running late. While the banter continued, Brendan pulled out his camera and started snapping candid shots.

"What's this Brendan, before and after shots?"

"Yea," Brendan retorted while maneuvering for another shot, "I want to be able to show the authorities what you looked like before this trip."

Raucous laughter, obscene gestures toward the camera, affected poses—the trolley nearly ran over us when it careened to a stop behind us.

As usual, the trolley was crowded, so we separated as necessary for the half-hour ride. The little trolley bumped along merrily, hugging the coastline all the way north to its terminus in Oostende.

Oostende was the main railway terminal for this portion of Belgium. Once we reached the train station, we would evaluate which trains we could connect with, and how much it cost to travel to various destinations. Then we would decide where we were going.

We also wanted to check out the prices for rental cars, but we already had a pretty good idea that the trains would be the better deal.

The trolley wove its way through downtown Oostende, between the centuries-old churches and buildings and across the brick roadways and alleys until it finally bumped to a halt in its terminal. We hustled outside, across the drawbridge overhead rows of beautifully polished wooden boats, and quickly located a rental car agency.

The train station was right beside the trolley terminal, but we wanted to check out the rental cars first.

It was the salesman in the Avis office who first dubbed us "The Magnificent Seven." He was, apparently, an ardent fan of old American movies.

But, he was right. We were, at least in our minds, the magnificent seven. Six of us were USAir pilots, and Ray Byrum rounded out the group with a Delta presence.

We were, by crews: Artis Galbreath and Brendan Kelly, Todd Kelly and Ray Byrum, Dave Richards and myself. LTC Pick Freeman provided the seventh element—the adult supervision, so to speak.

It rapidly became obvious that renting two small cars, or one mini-bus for the seven of us, was not going to be a cost effective solution to our transportation needs. We liked the idea of being able to arrange our own schedule, but we were leery about accepting responsibility for the vehicles in whatever foreign cities we decided to visit.

After we told him we wouldn't be able to rent one of his vehicles, the Avis salesman told us about a couple of other rental car agencies in town. We decided to split up and execute a rapid fact-finding tour.

Dave, Brendan and I headed for downtown to check out one of the new rental car agencies, while Todd and Ray checked out the other. Pick and Artis headed back for the train station to check out railway ticket prices.

When we rendezvoused again, the train was the hands-down winner for cost and convenience. They had a special tourist's package which allowed us to go anywhere in Belgium or Holland for an especially attractive rate. We decided to take that deal and go as far as we could.

That, is exactly how the magnificent seven found ourselves on the railway to Amsterdam.

Chapter 11

Our route to Amsterdam connected through the world's largest seaport, Rotterdam. Then it sliced through the low lands of the Netherlands past picture-perfect Dutch houses and windmills, finally depositing us in the Amsterdam train station in the late afternoon.

A conductor came by and ran us out of the the first-class rail car where we originally tried to ride. He smiled graciously as we tried to pass ourselves off as ignorant tourists. He knew we weren't *that* ignorant.

Shortly after settling into the more meager accommodations of the coach-class car, we were visited by a steward pushing a little cart filled with refreshments. A contest quickly ensued to see who had enough of the correct currency to purchase the first round of exotic refreshments.

Soon after the steward left us, an elegantly-dressed elderly lady passed through our car, no doubt enroute to the facilities located between the rail cars.

As soon as she was out of earshot range, Brendan said, "She reminds me of my grandmother's best friend, Gertrude."

"Oh?" Ray replied, instantly recognizing a lead-in.

Brendan did a fair job of keeping a straight face as he continued, "Did I ever tell you the story about

Gertrude and her fiance, Delbert, taking a buggy ride?"

Always a good straight man, Ray played his part by saying, "Why no Brendan, I don't believe we've heard that story."

And, for the next several miles, Brendan told us his tale of Gertrude and Delbert.

Delbert was giving Gertrude a ride home from their Texas church's chilli supper, when a tremendous west-Texas thunderstorm marched across their path. The thunderstorm's long tongues of lightning and rolls of ominous thunder threatened to overtake them before they reached Gertrude's home.

The thunderstorm was not the least of Gertrude's worries, however. It seems that the chilli at the church supper had not settled very well with Gertrude. She was experiencing a considerable amount of intestinal distress, but she dared not relieve it for fear of affecting Delbert's opinion of her as a lady.

Finally, as the thunderstorm drew almost even with them, Gertrude decided that she couldn't stand the strain any longer. She had been timing the intervals between the lightning flashes and the thunder.

She felt like she could probably time it so that she could raise a cheek and roll a little relief off the downwind corner of the buggy seat from Delbert. If timed properly, the tell-tale sounds of the affair would be muffled by the thunder, and the wind would take care of any lingering aromas.

Just when she thought she was going to burst, a gigantic bolt of lightning struck just behind the buggy. Gertrude immediately leaned over toward Delbert, feigning fear, but actually executing her plan with perfect timing.

As the last echoes of thunder died away on the wind, Gertrude said, "There certainly is a lot of wind in this storm, isn't there Delbert?"

Delbert replied, "There certainly is dear."

Reassured, Gertrude continued, "That thunder is awfully loud, isn't it Delbert?"

Delbert nodded as he replied, "It certainly is dear."

Confident now, Gertrude pressed, "That last bolt of lightning was certainly close, wasn't it Delbert?"

Delbert snorted and said, "It sure was, by the smell of things... I'd say it probably struck an outhouse close by."

Our laughter aroused the curiosity of the other passengers in the car. The seats in front of us and behind us were soon filled with eager eavesdroppers.

Dave said, "That reminds me of a story that a flight attendant named Gail told me, about a little country girl visiting her grandmother for the first time in the big city."

The steward passed back through the car at about this time, and Dave's story was temporarily delayed while we refilled our cups. Soon though, Dave was telling us his story about little country Annie.

Annie's grandmother, Granny, was delighted to have Annie visit her in the big city. To celebrate her visit, Granny bought Annie several presents, including a new dress and several new toys.

Of all the toys, Annie's favorite was a tiny watch. Annie loved to listen to it tick, and often wound it several times during the day to make certain that it didn't stop ticking.

Another favorite present of Annie's was a tiny bottle of little girls' perfume that Granny bought for her. It smelled sweet and fresh, like spring flowers.

One day, Granny decided to take Annie around the town with her, showing her off to all her friends.

To Granny's amusement, she soon discovered that Annie was not too well versed in the finer points of showing off her new presents.

Annie would tug at the skirts of total strangers on the sidewalk and ask, "Do you want to hear my new watch tick? Do you want to smell my new perfume?"

After returning the bemused smiles of the strangers, Granny was quick to instruct Annie in the etiquette of displaying her gifts.

Granny said, "Try not to be so forward dear. If a person inquires about your perfume or your watch, it's perfectly acceptable to tell them about it. Otherwise, it's not polite for you to bring up the subject."

Annie nodded her understanding, and Granny led her into an office building for the elevator ride up to meet Granny's friends.

Two well-dressed gentlemen joined them in the elevator for the long ride to the top of the building.

Annie tolerated their lack of attention for as long as she could, then she boldly tugged at the pants leg of the gentleman standing closest to her.

He smiled down at her and said, "Yes honey?"

Annie smiled back and said, "If you should hear anything or smell anything... that's me."

Things calmed down for a little while, Pick tried to read a paperback book while the rest of us enjoyed the scenery outside.

Just to see if Pick was paying attention to us, Artist said, "Look at all those girls—it must be a nude beach!"

Pick knew he was being baited so he didn't even look up.

Todd said, "That must be a good book Pick."

Pick replied, "As Groucho always said, 'Outside of a dog, a book is man's best friend. Inside of a dog, it's too dark to read.'"

Our chuckles were just enthusiastic enough to coax Pick to put down his book. He dog-eared a corner, folded the cover closed, and said, "I ran into an old gal the other day that I hadn't seen in a long time—ol' Ethel."

Artis kept him going with, "Where did you know Ethel from Pick?"

"Oh, I used to fly with Ethel all the time on the F-

28, but she retired a couple of years ago.

"I asked her how she'd been doing and if she'd seen any of the old gang, and she told me about running into ol' Stan the Man Cherrybomber."

I asked, "Who was Stan the Man Cherrybomber?"

Pick laughed first, then said, "Stan was a guy that should have been in all of your books, Sherm. He was absolutely crazy.

"By the time Stan retired back in '85, he had pulled just about every trick in the book. Ethel said he was getting a bit senile though."

Todd said, "If he retired in '85 he shouldn't be senile—he should only be about 67 years old."

Pick said, "Stan was fairly senile when he was 37. Ethel always said that most pilots considered baloney wax to be the best form of birth control, but Stan's was his personality... and his lay over clothes.

"Of course she was also fond of saying that the only difference between pilots and government bonds was that bonds would eventually mature."

Ray asked, "Did Ethel say that ol' Stan the Man remembered flying with her?"

Pick said, "She said that he recognized her sitting on the park bench, but she didn't really think that he could remember where he knew her from."

Dave asked, "What made her think that?"

Pick said, "She said that ol' Stan walked up to her, smiled real big, and said, 'I bet you can't guess how old I am.'

"Ethel said, 'Sure I can, drop your pants.'

"So, ol' Stan dropped his trousers right there in the park, and Ethel said, 'You're sixty-seven years old.'

"Stan said, 'How can you tell?'

"Ethel said, 'You told me so yesterday—*fool!*"

Even our Dutch eavesdroppers got a good laugh out of that one. But then, Pick Freeman knows how to tell a joke.

When things calmed down a little, Artis asked Pick, "Was he really that bad?"

Pick chuckled and said, "No, not really. He just liked to pull crazy stunts. One morning I saw him step out of the elevator in the hotel in Miami with a pair of panties strung over his head.

"When the lead flight attendant saw him and gasped, he just smiled and said, 'You know Lyn, I just couldn't get you off my mind last night!'"

The ride must have lasted four or five hours, but it seemed like about twenty minutes. The whole car was still laughing at Pick when the train pulled into the station at Amsterdam.

Chapter 12

Centuries ago, flood control and irrigation were the primary motivators for throwing a dam across the mouth of the Amstel river. Little did those early Dutch engineers suspect that a town of such world-wide fame would grow up around their little dam.

I wonder if those engineers could visit the train station in Amsterdam today, if they would still build the dam.

If I sound a bit cynical about the Amsterdam train station, it's only because I've been there. This is the first encounter that many visitors to the city have with the city's less desirable elements.

We hadn't even made it out of the station before a spaced-out junkie was trying to sell us drugs. The sidewalks outside the station were crowded with pan-handlers, all begging for money for drugs. We tried our best to ignore them, and the vulgar epithets they tossed after us.

We knew that we wouldn't get far without local currency, so all seven of us headed for the currency exchange office in the lowest level of the station to cash in some traveler's checks for Dutch gilders.

Two of the local seedy characters followed us into the currency exchange office. They wore long dirty coats over their grimy clothes. They hadn't shaved in at least a week, nor bathed in two.

The clerk behind the counter in the exchange

office handed us the usual currency exchange papers. One of the locals stood so close to me at the counter as I filled out my sheet, that I could smell his stinking breath. He lifted his arm to run his fingers through his dirty, oily hair, and his under-arm odor was nearly enough to make me nauseous.

"Why are they wearing those long coats in this spring weather?" I wondered. "They probably sleep in them on park benches at night," I decided.

My antagonist was standing as close as he could get, smoking a cigarette. I felt like he was purposefully exhaling his smoke in my direction. It finally occurred to me, that he might be deliberately trying to distract me to the point that I would forget about my bag sitting on the counter.

I made a show of looking at the guy as I passed my arm through the loop handles of my bag, then spread my feet to show him that I wasn't going to slide down the counter—away from him and my bag.

When I turned my attention back to my paperwork, he gave up acting like he was waiting in line and moved to the back of the room.

One by one, all seven of us picked up our gilders at the window and stepped outside the cramped office to wait for the others. Dave and I were standing around discussing the locals, when Brendan came out of the office and said, "Very funny Richards—now let's have it."

I could tell Dave was sincere when he answered,

"What are you talking about?"

Brendan said, "I'm talking about my camera Dave—where is it?"

Dave knew immediately what had happened. He shook his head and said, "I wouldn't kid you about a thing like this Brendan—if I'd seen your camera I would have picked it up for you, but I didn't realize you had taken it off your shoulder."

Brendan could tell he was telling the truth. He said, "I had to take it off my shoulder to get into my bag for my traveler's checks. Nobody else picked it up?"

A sickly sinking feeling passed through all of us. We hadn't been in Amsterdam for fifteen minutes and we were already victims of their pervasive drug culture.

Brendan's camera was a very nice 35mm model which he had paid a good deal of money for. That was bad enough, but even worse was the feeling that we hadn't looked out for each other—that we had acted like seven individuals when we recognized the threat instead of like a cohesive group. It made us all sick.

Pick, Brendan and I went upstairs to the police office to fill out a theft report. The other four stayed downstairs together, watching each other's backs.

The Dutch police acted like we were wasting their time. They told us to take a seat in the waiting room and they would get to us when they got a chance.

After waiting a while, I told Pick and Brendan that

I would go downstairs and tell the others what was happening. They could at least start making some calls, looking for lodging for the night.

On my way downstairs I passed a phone booth. I thought I would pop in quickly and phone back to Koksijde—just a quick message for Gary to let him know where we were.

I had just dropped my coins in the phone slot when a group of hoods started pounding on the door of the phone booth. They were making so much noise that I knew I couldn't be heard on the phone. So, I hung up, retrieved me coins, and opened the door.

A tirade of vulgar gestures and taunts greeted me as I emerged. Knowing that if I decked one of them the others would pounce on me and relieve me of my wallet—I ignored their taunts and made for the stair-way.

Downstairs, I found the guys and told gave them the update. They agreed to start calling hotels. I headed back for the police station, giving the wild-eyed phone booth crowd a wide berth.

We were debriefed later that it was a good thing that I was inside the station when I encountered these guys. They're a little less daring inside the building where their escape can be hampered.

We learned later about another group of our guys that was visiting the city, when one of them stopped on the sidewalk to retie his shoelace. When he straightened up, a guy stuck a switchblade under his chin. Even though his buddies were just a few paces away, the American was afraid to call out for help. He handed over his money, and watched his assailant disappear down an alley.

Needless to say, by the time the police report was completed and we were all together again, we were seriously considering getting right back on a train out of Amsterdam.

But, we were tired, and no one really felt like riding a train anymore, so we shuffled outside to catch a trolley downtown. Being the trooper he is, Brendan shook off his disappointment and resumed his normal role of court jester. That's the type of guy Brendan is—always thinking of the rest of the group before himself.

Downtown, all the major hotels were incredibly expensive, even with our airline employee discounts. The high prices of everything in Europe were quickly turning the entire trip into an America-appreciation sortie.

We had just about decided to suck it up and stay at one of the major chain hotels, when Artis saved us. He had noticed a less-expensive looking hostel around the corner from the major establishments, so he asked us to hold off for a minute until he checked their prices.

Still suffering from hotel sticker-shock, we readily agreed. Artis walked over to the Westropa Hotel, and found us a deal. We got two single rooms, one double, and one suite for three of us—all for about the same price as one room at the "airline hotel." And, the price included breakfast!

Ray and Todd took the double room. Being senior, Pick naturally rated one of the single rooms. We tossed a coin for the other single, and Brendan won. Dave and Artist and I got the suite.

We cleaned up a little bit then walked to the restaurant strip for dinner. We had some fantastic steaks at the Argentina steak house. After dinner we made the rounds of the various pubs along the alley, each featuring its own national flavor.

We made our way up one side of the alley and down the other, listening to jazz and blues and rock & roll until we were too tired to keep going.

Our voice of experience, Pick, wisely counseled us to call it a night so that we could get up in time to take advantage of our complimentary breakfast, and enjoy the cultural offerings of the city on the morrow.

We were tired, so we acquiesced.

The last thing I remember of the day was turning off our lights, and sinking into my over-soft pillow.

Dave, doing his best Waltons impersonation, whispered across the dark room, just loud enough for Artis to hear, "Goodnight Sherm boy."

I answered, "Goodnight Davey boy."

Artis let it go for about ten seconds, then finally grunted, "Shoooot"—or something like that—and went to sleep.

Chapter 13

I can't remember if the window in our room didn't have any curtains, or we just forgot to pull them closed. Either way, the result was that the sun came pouring into our suite the next morning, slashing across our faces like a wide-beamed laser.

My bed was closest to the window so I got hit first. It was just the advantage I needed to beat Dave and Artis to the shower.

We met the rest of the magnificent seven in the lowest level of the Westropa hotel, where we shared the complimentary breakfast.

Most of the breakfast talk centered around what we were going to do with our day. Brendan was big on going to the Van Gogh museum, which of course elicited the typical response from Dave, "I already know all about that...van go, Beep-Beep."

Todd mentioned taking the canal tour, while Pick wanted to do some diamond shopping. We decided that if we stopped packing in the waffles and got on the road that we could do it all—so we did.

The desk clerk at the Westropa agreed to let us check out of our rooms but leave our bags behind their counter while we explored the city. That freed us up to move fast.

We did some quick walking and sight-seeing, stopping occasionally to let Pick look at diamonds.

Our walking tour led us to the Van Gogh museum,

where Brendan sprung for the little hand-held radio that quietly told the story of each painting as you stood in front of it.

I've always admired Van Gogh's work, but I had a hunch that it would be more fun to let Brendan tell the story than to listen to it on the little white radio. I was right.

Brendan had a way of paraphrasing the Dutch broadcasts, which made each painting seem much more personal and meaningful to our little group. He knew just which information to extract from the broadcasts, and how to put his own spin on the information to make each painting tell a story.

When the tour was over, we were all glad that we had let Brendan tell us Van Gogh's life story through his works, rather than having each of us stand silently in front of the paintings with a tiny white radio against our ears.

Our train schedule told us that we were running out of time. If we were going to get in the canal tour, we would have to go directly from the boat tour office to the train station in order to make our connections back to Oostende.

When you see the canal tour boats for the first time you're immediately reminded of the James Bond movie where you probably first saw them. They are sleek, all-glass enclosed models with terrific views, and they serve refreshments!

I'm not sure what brought up the topic of lawyers,

but somehow as we cruised up and down the famous canals of Amsterdam, the guys started telling every lawyer joke they had ever heard.

I won't repeat them all here, mostly because I know you've already heard them all, but there was one that I thought was pretty good.

Artis told the tale about a Mississippi lawyer who had been hired by an insurance company to try to keep the company from having to pay off on all of its claims.

One of the claims involved a client who had passed away while holding a sizeable life insurance policy. The insurance company dispatched their lawyer to see what he could do about getting out of the claim.

The lawyer's first stop was the office of the doctor who had signed the death certificate.

The lawyer's first question was, "When did you actually see the *alleged* body?"

The doctor replied, "I'm afraid that I didn't actually see the deceased gentleman's body—I was called away at the last moment before my examination of the body to deliver a baby. However, my partner did see the body, and I signed the death certificate based upon his findings."

"Why didn't your partner sign the death certificate if he was the one that saw the *alleged* body?"

"He was called out of town to perform emergency surgery before he had a chance to do the paperwork.

126

Since the state requires all death certificates to be signed within 24 hours, I simply signed the papers based upon my partner's findings."

"I see," snapped the attorney, "and I suppose then that you also did not actually perform an autopsy on the *alleged* body?"

"No, I didn't. As I've explained, my partner performed the autopsy, I did not see the body."

"Uh huh. Let me ask you this," the lawyer proceeded in his best trial voice, "did you attend the *alleged* deceased man's funeral?"

"Well, no sir, I did not. I was on duty in the emergency room at the hospital on the day that the gentleman's funeral services were conducted."

"I understand!" exclaimed the attorney as he slapped the doctor's desk—looking for all the world as if he had just discovered a well-disguised secret.

"You never saw the *alleged* person's body, nor did you perform an autopsy on the *alleged* body, nor did you attend the *alleged* dead man's funeral. Now the truth is, doctor, that you don't really know for certain whether the man is dead and buried or not... do you?"

The doctor nodded and said, "You know... you're right. I do know for certain that the man's brain is sitting in that jar over there on the shelf behind you, but for all I know, he could be running around Mississippi right now practicing law!"

We finished the canal tour just in time to catch the last train for Belgium. We talked about catching a train to Germany but decided against it. The German air traffic controllers were on strike, and we were afraid we would get stuck over there.

The train-to-Germany idea did spawn one announcement from Dave. He hooked us with the question, "Hey, did you guys see that article in the paper about East Germany and West Germany agreeing on where their new capital should be?"

We gave a collective head shake, and Dave said, "Paris!"

I remember one of the guys trying a sloppy joe out of a vending machine in one of the train stations on the way home. It was awful. We found out later it was made with horse meat.

We had planned to have dinner in Oostende before catching the trolley back down the coast to Koksijde, but Pick wisely walked across to the trolley station when we finally arrived in Oostende, and discovered that we had barely enough time to catch the last trolley home!

By pure stroke of luck, we hopped on the final trolley south for the evening, and finally completed our journey by stepping off at the last trolley stop in Koksijde.

It just so happened that the last trolley stop was right next to a restaurant, which happened to have a dinner special which was moderately priced. I've already said that moderately priced meals were at a premium in Europe, so we readily took advantage of the opportunity to share a final meal together.

The food and wine were great. We knew that starting tomorrow, we were going to be split up and sent all over Europe as individual crews.

But on that night, it was a perfect conclusion to the Amsterdam adventure of the magnificent seven.

Chapter 14

Friday morning dawned clear and bright on the Belgian coast. Dave and Terry and I joined the throngs of other crewmembers which were fighting for space on the morning bus.

The bus which ran from the Holiday Park to the airfield was designed to accommodate 30 passengers. We packed on about 65 guys with all their bags.

We had used Thursday as a catch-up day—washing our clothes from the Amsterdam trip and getting our mission briefings for today's adventure. We polished off the night before with more tennis and another spaghetti dinner, featuring Jim Young's cooking again. It was great, but we were all spring-loaded to the "let's go" position this morning.

Our mission was to fly from Koksijde, Belgium to Bitburg, Germany to pick up some cargo. We would then deliver the cargo to Bentwaters, England. Then we were scheduled to fly empty, no cargo, from Bentwaters all the way to Naples, Italy.

I didn't like the idea of flying all the way to Italy empty, so I called the load planners in Ramstein, Germany and asked if they really wanted us to do that. I was sort of hoping they would say, "No, what we really want you to do is fly to Moscow!"

But, no such luck. The controllers at Ramstein said that we were flying to Naples in order to pick up cargo for the following day, Saturday.

On Saturday, we were scheduled to fly from Naples, Italy to Rota, Spain. After dropping off the Naples cargo in Rota, we were supposed to pick up a load of cargo for Sigonella, Sicily. We were scheduled to fly all the way back to Sigonella on Saturday, arriving late in the evening.

On Sunday we were supposed to fly from Sigonella, Sicily back to Naples, Italy. After dropping off cargo in Naples, we were scheduled to continue to Ramstein, Germany, where we would spend Sunday night.

On Monday we were scheduled to fly from Ramstein to Koksijde, where our mission would end.

That was a lot of flying in a C-130. To make the mission even more interesting, we were supposed to follow another Air Guard crew from a different base around the entire course. They were scheduled to takeoff from each location exactly five minutes prior to us. They were flying a newer, faster model of the C-130.

The flight to Bitburg only took one hour. The sky was a clear blue, the fields were lush greens, and we were considering ourselves lucky to be there.

The flight across the English channel to Bentwaters took a little longer—1.6 hours.

It didn't take long at all for Ned and Tommy to unload our airplane in Bentwaters. Since we were flying empty to Naples, we could have easily taken off before the other crew if we had wanted to.

We nearly departed before them at Bitburg, which was a source of great irritation to the aircraft commander of the other crew.

He was a rather diminutive fellow named Ester. His crew seemed to be having a difficult time getting along with him also—I think he may have suffered from a slight Napoleonic complex.

My engineer, Jim Young, was standing next to me in the operations building at Bentwaters when Ester made an admonishing remark in my direction that he *expected* us to wait until he was airborne before we taxied for takeoff.

I think Ester was still within earshot when Jim "whispered" his thoughts on the matter. Of course, I sometimes think that Jim learned to whisper in a sawmill.

Jim leaned toward me and said, "He's so short, I'll bet that when he farts he blows sand in his shoes!"

Ester kept going as if he hadn't heard it.

The flight from Bentwaters to Naples took 3.4 hours. A good portion of the flight was over an undercast, so there wasn't much sightseeing to be done.

Old jokes started flying over the intercom, just enough to keep everybody awake and interested in what was going on.

Jim Young started it off by asking if we had heard of that new Iraqi beer named after Saddam Hussein's missiles—Scud Light? No matter how many you have, they never hit the spot.

Ned and Tommy didn't have much to do in the back since we were sans-cargo, so Ned told us the story about God awarding sex lives to the creatures of the earth.

According to Ned, God originally granted man only 20 years of active sex life. Of course, man wanted to extend that if he could.

God also awarded the horse with a 20 year sex life, but the horse wasn't as concerned about it as man was. As a matter of fact, the horse was willing to give up half of its sex life if God would just remove the horn sticking out of the middle of its forehead.

The lion was also initially awarded a 20 year sex life, but he really cared more about being the king of the jungle than having 20 years of sexual activity. The lion petitioned God to take away half of his sex life, and award him a regal mane so he could be king.

The donkey also initially had a 20 year sex life, but he thought it was way too much. The donkey offered to trade half its sex life for a nice long tail to swat insects from its hindquarters.

God decided to grant everyone's wishes. He took half the animals' sex lives, granted their wishes, and gave the extra years of sexual activity to man.

Which, according to Ned, explains why to this day, man has 20 years of normal sex life, 10 years of horsing around, 10 years of lying about it, and 10 more years of making an ass of himself!

Dave didn't seem to enjoy Ned's joke as much as the rest of us, but then, he and the loadmasters were having a running practical-joke war for the entire trip. Dave might have still been a little unhappy about the tie-down chains that Ned and Tommy put in his helmet bag at the beginning of the trip.

When Ned finished telling his joke, Dave said, "That sounds like the kind of story a loadmaster would tell."

Tommy retorted for the loads with, "Hey Dave, how many pilots does it take to screw in a light bulb?"

When Dave refused to answer, Tommy said, "One—he just holds it up and waits for the world to revolve around him."

Dave responded with, "How many loadmasters does it take to change a light bulb?"

He didn't give them a chance to come up with anything cute to say before he answered his own question, "Five—one to do it, and four to stand around and say, 'That ain't the way they told us to do it at tech school!'"

I thought I would break it up while the score was more or less even, so I threw one in too. I said, "Do any of you guys know how many cowboys it takes to change a light bulb?"

Dave grinned at me, and I answered myself, "Two—one to change it, and one to write a sad song about how much they miss the old light bulb."

It was dark now and Naples was socked in, so we

devoted ourselves to some intense study of the approach plate and terrain chart in preparation for our arrival at Naples.

The approach into Naples slides in between some fairly large hills, so we wanted to make sure we kept the airplane on course and at the proper altitudes.

Dave flew the approach into the airfield, and of course he did his usual superb job. It wasn't until we were leaving the next day, when we could clearly see the mountains around Naples, that we appreciated exactly how well he had done.

Before we left the airfield for our hotel, we briefed the Navy aerial-port people that we planned to return in the morning, upload the Rota cargo, fly to Rota and then Sigonella on Saturday, and then drop back into Naples on our way to Ramstein on Sunday.

We were surprised to learn that they wouldn't be open on Sunday! Whoever set up our mission, failed to check the operational hours of Naples. The local Italians who drive the trucks and forklifts and loaders, always take Sunday off. Since the workers take the day off, the Navy supervisory personnel also take the day off.

It occurred to me that this might present an opportunity to get a little bit better crew-rest deal for my crew tomorrow. But, tonight, everyone was very tired. We clambered into the buses with our bags and headed downtown for our hotel.

Neither Ester's crew nor my own had eaten dinner.

So, upon our arrival at the hotel, we agreed that both crews would meet in twenty minutes to go to a restaurant together.

The most memorable part of the next twenty minutes was the shower. The Italian showers were almost exactly the size of a coffin. A tiny shower curtain was provided to keep the water from spraying directly out onto the bathroom floor, but since none of the drains worked the water wound up on the floor anyway.

The best way that I can imagine to duplicate that shower, would be to wrap yourself up in a plastic sheet and spray water on your head while standing in a bucket.

But, after the hard day's flying, the water did feel good, and most of the crew seemed semi-revived by the time we met in the lobby twenty minutes later.

We asked the desk clerk for a restaurant recommendation, and then for a guess as to how many dollars we needed to exchange for lire to pay for the meal. He recommended buying an amount of lire which was equivalent to twenty-seven U.S. dollars.

As soon as everyone had their lire we jumped into some waiting taxicabs which hustled us down the hill to the restaurant recommended by the desk clerk.

I have to admit, the food was excellent. Nobody ever saw a menu. We just took our seats at a long table, and the waiters started bringing food. They brought dish after dish of great Italian food for the next couple of hours, along with enough vino to keep the food washed down.

It was great, but we were also a little suspicious about the price. Sure enough, when the waiter brought the bill, it worked out to exactly the equivalent of $27, with tip. We had the definite impression that we had just dined in the desk clerk's brother-in-law's restaurant, but nobody cared.

It was a fine evening.

Chapter 15

Saturday morning came a little too early for most of us. In spite of the terrific meal the night before, most of the crewmembers could have stood a bit more rest before clambering back into the airplanes.

The bus ride between the hotel and the airport was even more exciting in the daytime. We could see everything that we almost hit, or that almost hit us. It's true what they say about the Italian cars—the most important component is the horn. Our driver had an exceptionally loud horn—riding with him was definitely worth an E ticket at Disneyworld.

After dropping our bags off at the airplane, Dave and Terry strolled to operations and started flight planning. Jim, Ned and Tommy stayed at the airplane to load and preflight. I got on the telephone.

I called the aerial port people at Rota and explained to them that we were scheduled to pickup cargo for Sigonella and Naples, but Naples would be closed on Sunday. I was hoping they would not have two C-130 loads of cargo for Sigonella.

If so, I was planning to let Ester take the Sigonella cargo to Sicily, while my crew either crew rested in Rota, or returned directly to Naples. The flight from Rota to Sigonella was nearly two hours longer than the flight from Rota to Naples. That would give my crew two more hours of rest in Naples tonight, and two hours more hours of rest in Ramstein tomorrow.

The Rota aerial port supervisor told me that they had much more than two C-130 loads for Sigonella. He thanked me for telling him that Naples would not be open, and said that he would make certain that both of our airplanes were loaded exclusively with cargo bound for Sigonella. A good idea, but it looked like I wasn't going to get the extra four hours of crew rest for my guys.

We made an on-time takeoff from Naples, chasing Ester's crew west across the Mediterranean. The guys were very tired. There was very little talk on the intercom.

I've done quite a bit of international flying in C-141 Starlifters, C-5 Galaxies, and the C-130 Hercules. We had longer crew duty days in the C-5, but overall they were not as tiresome as flying the C-130.

The Hercules cockpit has more noise and vibration, and usually the temperature and pressurization is more difficult to control. The workload on the pilots in greater since the 1958 vintage autopilot can not be coupled to any navigation functions, and of course there are no bunkrooms on a Hercules.

I could definitely tell that the heat, noise, vibration and temperature/pressurization deviations were taking its toll on my crew.

Naples is on the west coast of Italy, about halfway up the laces of the boot if that's the way you picture that country. Rota is on the west coast of Spain, close to where Columbus set sail in 1492. The flight took

3.6 hours.

When we arrived in Rota, Dave, Terry and I busied ourselves with flight planning duties while Jim, Ned and Tommy prepared the airplane for the next leg to Sigonella. Everyone worked as quickly as they could, wanting to get to Sigonella as soon as possible.

Ester's crew was barely airborne as we cranked our engines. I taxied out to the steeply-sloped east end of the runway at Rota, and we departed for Sigonella.

We had just reached cruise altitude when Ned asked over the intercom, "Hey Sherm, are we going to Sigonella or Naples?"

"Sigonella, Ned."

"Well, this forward pallet has a Naples tag on it. I wonder why they put a Naples pallet on our plane?"

I answered, "They weren't supposed to put any Naples cargo on our airplane at all—we're supposed to only have Sigonella cargo."

Ned said, "Hold on a minute, I'll check the other pallets and see where they're going."

A couple of minutes later Ned called me back with the news—every pallet of cargo on our airplane was destined for Naples! Some of the cargo contained medical supplies—the type of stuff that we imagined Naples would like to have delivered today.

We got on the radio and called back to Rota aerial port, and they admitted that they had screwed up and loaded the wrong cargo on our ship. They told us to take it on to Sigonella, where somebody else could get it to Naples.

We couldn't see much sense in taking the Naples cargo to Sigonella, especially since we didn't have a single piece of Sigonella cargo on our plane. So, I got on the HF (High Frequency) radio and got a phone patch with the duty controller at Ramstein.

The controllers at Ramstein are supposed to control all of the airlift in Europe. Their call sign in "Phantom".

After numerous tries, I was finally able to get an HF operator at Incirlik Air Base, Turkey, to get me a phone patch with the Phantom duty controller.

I explained the situation to Phantom, along with my recommendation as to the best way to solve the problem, namely divert us into Naples. I've always found that to be the best way to deal with duty controllers—is to not ask them to do your thinking for you. Tell them the problem, then tell them the solution, and get their concurrence with whatever action you want to take.

The first duty controller at Phantom that I talked to was a young enlisted woman. She readily grasped the situation, but said that she would have to get approval from higher up before she could approve our divert. She asked me to call her back in 30 minutes.

One very positive aspect of this whole exercise

was that the crew came alive. The prospect of cutting the flight short by a couple of hours today and chopping two hours off the flight tomorrow to Ramstein, was enough to perk everyone up.

Our route of flight east over the Mediterranean was the same for most of the route, whether we were going to Naples or to Sigonella. It was late in the flight before the route turned south, away from Naples, and angled down the coast of Italy for the island of Sicily.

I wanted to get permission to divert to Naples early, before we came to the southerly turn.

We used the thirty minutes to check the weather and NOTAMs at Naples. Everything was fine. No reason not to divert into Naples.

When I called the Incirlik controller back in a half hour, she knew exactly what I wanted. She immediately arranged the phone patch with Phantom.

This time, I got a male sergeant on the phone, and he seemed incapable of grasping the situation. He listened to my arguments, then said, "Permission to divert denied—stay on the original schedule. Continue to Sigonella."

It couldn't be more plain than that. The female HF operator at Incirlik terminated the phone patch, and I could hear the incredulity in her voice as she signed off. She couldn't believe it either.

In our cockpit, the mood changed from jovial to incredulous. The Phantom controller hadn't offered a single good reason for refusing our divert, it was becoming more obvious by the second that he simply didn't want to do the work involved in changing our itinerary.

The more we talked about the waste of what we were doing, the more angry I became. I finally decided to try another phone patch, hoping that I might get a different controller on the line.

The Incirlik controller seemed glad to hear from us again. She had a definite note of relief in her voice when she told me to stand by for a phone patch with Phantom. She didn't want us to give up either.

When the phone patch went through, I got the same sergeant on the phone. I asked him point blank exactly what their plan was for our medical supplies and the other Naples cargo. He told us that another crew was standing by at Sigonella, waiting to take our airplane to Naples.

Now, the rationale for this argument was that Phantom needed to get that ghost crew at Sigonella in place at Naples, and have us in place at Sigonella. This would have been a pretty good argument if we had been an active duty crew. Active duty crews often fly out of a "stage" like that—swapping the airplanes around between crews as necessary to do whatever mission needs to be done.

But, we were an Air Guard crew, and *nobody* flys our airplanes but *us*!

So, the sergeant's argument was quickly held up

in front of him for exactly what it was—a bare faced lie. He was simply hoping that we would proceed to Sigonella to meet up with this ghost crew. When no crew materialized, we would go into crew rest as scheduled, and sometime in the future, somebody would get the cargo to Naples. He didn't care when.

After I told the sergeant that we were flying a Guard airplane, he immediately changed his story. This time he said that what he meant to say was that another crew *and airplane* were standing by at Sigonella, ready to transfer our load to their airplane, and carry it to Naples.

He went on to say that Sigonella had a heavy load of cargo for Ramstein that we needed to carry on Sunday, so we couldn't divert into Naples or that cargo would be abandoned. He said that the "ghost" crew at Sigonella now, couldn't carry the cargo to Ramstein, because it would "take them out of their flow."

We knew by now that this guy was making these stories up as fast as he could think of them! It was making me mad. We didn't have a very good phone patch, but I still tried to make my case with a rather long-winded explanation of why the best thing for everyone would be for my crew to divert into Naples.

When I finished, the controller said that he hadn't been able to hear all of my transmission, so the operator at Incirlik repeated my argument for me. The tone in her voice as she spoke definitely said,

"Look you lazy idiot, get up off your fat duff and change the paperwork to show that these guys diverted into Naples!"

When she was finished, the Phantom controller still wouldn't approve the divert. He still maintained that another C-130 was standing by in Sigonella to pick up our load.

He also said, that the Naples aerial port office closed at 1700 local on Saturdays—nearly two hours before we could arrive there. He said that if we did divert into Naples, there would be no one to unload or accept our cargo, and nobody to refuel us for our flight out on Sunday.

I thanked the Incirlik controller very sincerely for her efforts, and terminated the phone patch.

Boy, we were mad. There's nothing worse than feeling like the people out in the field who know what's going on are being stifled by some lazy armchair quarterback in a dark room far away.

I couldn't just let it lay. We were rapidly closing in on the southerly turn, where the whole thing would become a moot point.

First, I called the Incirlik controller back and asked her for a phone patch with the Sigonella command post. She was practically jumping up and down with glee—she didn't want to give up either.

The Sigonella controllers said that they didn't know anything about any crew that was supposed to carry our cargo to Naples. Thanking them, I asked our friendly Incirlik controller for a phone patch with the Sigonella aerial port office. They said that they only had one pallet of cargo for Ramstein.

Ester was fully loaded with Sigonella cargo so he definitely had to proceed there—he could carry their one pallet of cargo to Ramstein tomorrow.

The Incirlik HF controller was ecstatic. She had to listen in on every phone patch, and she *liked* what she was hearing. As soon as I told her to terminate the phone patch with Sigonella and give me another one with Naples aerial port, she practically "Whooped" over the radio, "Yes *sir!*"

Aerial port at Naples said that they would be open until 7 pm local time—not 1700 (5 pm local). They also said that they could refuel us as long as we were ready to take on fuel by 15 minutes until 7. They said that their Italian refueler always quit promptly at 7 pm, but that he could refuel us in 15 minutes.

I finished the phone patch with the Naples aerial port people by saying, "Don't let anyone go home early—we'll be there at 1830 local!"

When I told the Incirlik controller to terminate the phone patch with Naples, I didn't even have to tell her what I wanted next. She said, "Yes sir, standby for phone patch with Phantom!"

This time, the original female controller I had talked to at Phantom picked up the phone. We were rapidly approaching the turn point, so I quickly described the situation to her. I shot down every one of the lies that her cohort had given us earlier, and I told her the true story about the operational times at Naples. I ended by telling her I needed an answer within one minute.

She asked me to standby for 30 seconds while she consulted with her senior controller, then the call came back, "Divert approved! We were directed to divert to Naples!"

The hoots and hollers in my airplane were intense, but short lived. We were going to be awfully busy for the rest of this flight if we were going to pull it off.

First we had to get a new clearance from the air traffic controllers, allowing us to divert into Naples. Everything we had done in preparation for our arrival at Sigonella had to be changed—and we were rapidly running out of time. Terry said that his best guess was that we would arrive at 1840.

We were determined to beat that. We kept the airplane at max speed throughout the descent and arrival, and arranged a special clearance to land straight-in, instead of flying around the airfield to land the opposite direction.

Somewhere in the midst of the arrival I felt like I had a few spare seconds, so I called the Incirlik controller back to offer her a proper thanks for her help. She had gone off duty, but a friend of hers had taken her place. Her friend gave me her name and address, and I sent her a thank you card and a copy of my last humor book.

I never heard back from her. I hope she got the card, she sure did a great job for us that day.

The mood in our airplane was jubilant—everyone working together, trying to help each other out, trying to cover for each other. We felt like a defensive football team that had just intercepted a pass.

We beat all the C-130 speed records for an arrival into Naples. We had the props stopped and the chocks in at 1825. Ned and Tommy practically threw the pallets off the back onto the waiting loaders, and Jim Young filled our fuel tanks as quickly as the truck could pump.

By 7 pm the Italians were on their way out the airfield gate, and we were waiting for our bus back downtown to the same hotel we had stayed in Friday night.

In spite of the 3.6 hour flight to Rota and the 3.8 hour flight back, we felt better now than we did when we started the day.

An hour later, I remember unwrapping the shower curtain from around my shoulders and stepping out onto the flooded bathroom floor, and thinking "Ester ought to just be touching down about now in Sigonella".

The crew was in high spirits as we departed the hotel for our favorite Naples restaurant.

Chapter 16

I suppose we could have delayed our departure on Sunday morning for a couple of hours, but by now the crew was back in the "grand adventure" mode. So, we elected to leave early and take our extra crew rest time in Ramstein instead of Naples.

As Jim put it, "Germany is so nice in the spring."

I agreed. Germany has always generated considerably more history than they could absorb locally—which is one of the reasons I like it so much.

As usual, the Phantom controllers gave us a little grief about arriving at Ramstein two hours early—they said the ramp would be too crowded to park us at that time. But, we had plenty of fuel, so we decided to go anyway. If the ramp was too saturated for us to land we would just do a little training in the instrument pattern until the ramp opened up.

Dave was in a particularly good mood this mother's day. He definitely came out with the most one-liners of the day. He flew the leg to Ramstein.

When the Naples tower controller cleared us for takeoff Dave called for the line-up checklist, then said, "Okay guys, remember, we're going to break *ground*, and fly in the *wind*."

We were empty so the airplane was light, and we took off at sea level on a relatively cool day. Dave pulled the Herk's nose way up and let her claw her way skyward away from Naples.

Dave noticed Jim and I exchanging glances over the steep climb angle. He winked and said, "Well boys, she's climbing like a homesick angle today, ain't she?"

We couldn't help but laugh.

Terry pressed his mike switch and said, "We need a turn to the north to intercept that outbound radial on the SID (Standard Instrument Departure)."

Dave said, "Okay, I'll hang a right turn to zero-one-zero and let her eat."

The weather was clear all the way past the French alps. Everyone shot up at least one roll of film on those beautiful snow-capped peaks.

Shortly after passing the alps an overcast layer of clouds moved in, obscuring the gorgeous checkerboard of green, yellow and brown fields below.

We started talking about the possible training flight that awaited us upon our arrival at Ramstein, and that spawned a discussion of old flight school stories.

I told one of my favorites, as told to me by one of my favorite captains, Terry Barber. Terry told me about a training flight he was on one day when he was a Navy instructor pilot in T-2 Buckeyes.

He said that he had a Marine student in the back cockpit of the tandem-seat trainer, and since it was an instrument training flight, the student was "under the bag." Under the bag meant that the Marine's canopy was completely covered with cloth, making it impossible for him to see outside.

The plan was for the Marine to fly the airplane down the instrument approach path, then turn it over to Terry to land when they reached the decision height on the approach.

The Marine did everything fairly well, except for one thing—he forgot to put the landing gear down. When he selected final flaps, the landing gear warning horn started blaring in their headsets, and a bright red light on the instrument console started flashing in their faces.

Terry couldn't believe that the student was able to completely ignore the landing gear warnings. But, he did—blithely continuing his descent down the glide slope, completely oblivious to the red "Wheels" light and the horn.

Finally, Terry pressed his intercom button and said, "Isn't that flashing red light and that horn distracting?"

The Marine answered, "A little sir, but I can hack it."

We were cruising at FL 200, but we started picking up some light turbulence as we neared the French border. Thinking it might be smoother at a lower altitude, I asked the French controller if he had any smoother altitudes available.

There was a deliberate note of disdain in the Frenchman's voice as he answered, "I've had no complaints at FL 180."

The controller was deliberately trying to be rude with his tone of voice, which didn't sit well with Dave. Dave pressed his mike button and said, "Does that mean there were no complaints, or no survivors?"

When we arrived over Ramstein it was obvious that Phantom had lied to us again, the ramp was almost completely empty, with no scheduled arrivals except us. We received permission from the command post to land immediately, so we told the German controllers that we would be making a full stop landing.

Ned and Tommy immediately started needling Dave about his upcoming landing. They were trying to make him so nervous about it that he would prang it on.

Dave answered their taunts by saying, "My theory of landing is, any landing that you can *walk* away from is a *good* landing."

Jim entered the fray by interjecting, "Why don't you try for one of those *great* landings that we can *taxi* away from."

I have to hand it to him, Dave didn't choke. He greased that 130 onto the runway so smooth that I was tempted to call tower and ask if we were down.

Even Ned in the back said, "Well I'll be... the boy set her down that time like a butterfly with sore feet."

Three hours after we lifted off in Naples, I pulled the condition levers back to stop, and the big Hamilton-Standard props started spinning down. We put the airplane to bed, then climbed onto the crew bus to head for billeting. The entire crew wound up staying in the same building—the building which also housed the billeting office.

Ned and Tommy said that they were going to spend their afternoon shopping at the base exchange, and Jim said he was going to enjoy his "chief's quarters". Terry said that he was going to do some reading in his room.

Dave and I told them to meet us at the club later if they wanted to get together for dinner, then Dave and I changed into our running clothes and headed out for a little exercise.

There is a path that runs through the woods on Ramstein Air Base. It's a narrow path, but still a beautiful place to run. I, and about a hundred other guys, slept beside that path one night in the early 80's when Germany was experiencing a rare heat wave. The BOQ (Bachelor Officer Quarters) rooms have no air-conditioning, so most of us escaped the oppressive heat of our rooms that night by carrying our blankets and pillows into the woods.

Dave and I ran through the woods, then down to the flight line, then back along the flight line toward the BOQ. We passed the BOQ office and continued to a recreation area which included a golf driving range.

Dave is an avid golfer. As such, he naturally has

a fairly large repertoire of golf stories.

To keep our minds off the run, Dave decided to tell me a golf story as we ran past the driving range.

His story was about a bride and groom, who unexpectedly have to leave their honeymoon accommodations and find a new hotel. The groom goes to extreme lengths to find a hotel adjacent to a golf course.

Finally, the bride asks, "Is there some sort of problem here that I should know about?"

The groom answers, "As a matter of fact, there is. I was afraid to tell you this before the marriage, because I was afraid you wouldn't go through with the ceremony. But, the truth is...I'm a golfaholic. I *have* to play golf, every day!"

The bride shrugs and says, "As long as it's true confession time, I suppose there's something that you should know about me... I'm a hooker."

The groom smiles and says, "No problem honey, I can help you fix that. Just stand with your left foot a little bit further forward than you usually do, and aim just a little bit to the right..."

Dave's story ate up a little time, which I was grateful for. He added to it slightly as we were running back toward the room. We were passing a Subaru car in the parking lot, when Dave said, "Hey Sherm, did you ever notice that Subaru backwards spells: U-R-A Bus."

We finished our run and got cleaned up for a good

German dinner. We were on our way out the front door when we ran into Ester's crew, on their way in to the billeting office. They didn't look like they were getting along very well.

Dave and I strolled over to the Officer's club where we found Jim Young waiting for us in the bar. He was enjoying a tall glass of weissen beer, a wheat-beer that the Germans serve with a slice of lemon in the glass.

Dave said, "That beer sure looks good."

Jim said, "You guys be careful with these German beers. A couple of these things will have you boys seeing double and feeling single."

He said that everyone else had made other dinner plans, so the three of us called for a cab and headed downtown to K-town (Kaiserslauten) for one of my favorite German restaurants.

A half hour later we were relaxing over a round of weissen beers, waiting for our Jagerschnitzels to be prepared, when the conversation turned to famous characters who had been part of the North Carolina Air National Guard.

Jim had dozens of stories, but my favorite was his tale about Colonel Sweetwater.

Lieutenant Colonel Frank Sweetwater was a pilot in my Guard unit back in the early 70's. Frank also flew for Eastern airlines.

According to Jim's story, Frank was out flying a C-130 mission one day, when the N1 compass system went out. That meant, that the only heading reference in the airplane that still worked, was the magnetic compass.

Normally, Frank would have landed at Pope Air Force Base and had the problem fixed, but he was running late. He was supposed to fly for Eastern that afternoon—he had left his Eastern uniform hanging in the locker room back at the squadron. His plan was to grab a quick shower, then hustle across the ramp and sign in for his Eastern trip.

Now, Charlotte was socked in with low clouds, and Frank's only heading reference was the erratic mag compass.

Not to be deterred, Frank asked for vectors to the final approach course, then successfully flew the instrument approach using only the mag compass for heading corrections.

Of course Frank wrote up the faulty N1 compass system, but he put it in the forms as if it had malfunctioned on the ground, as he taxied in after the flight.

Much to Frank's chagrin, the rest of his crew didn't play along with his plan. They were so impressed by his superior airmanship, that they wrote him up for a commendation, asking that he be awarded a D.F.C. (Distinguished Flying Cross) for the flight.

When Frank's superiors read the commendation paperwork and discovered what he had done, they were willing to do anything *but* give him a commendation.

Jim laughed and concluded his story by saying, "Frank left the Guard shortly after that. He was our only pilot to have a D.F.C. downgraded to a flight violation!"

We enjoyed a great meal in K-town, then caught a cab back to the base for a good, well-earned night's sleep.

Jim and I had no problem falling asleep and staying that way for the rest of the night.

Dave was not so lucky.

Chapter 17

I awoke the next morning to the unmistakable sound of someone trying to jimmy my door.

I lay there for a moment, gathering my bearings, and trying to imagine who might be trying to break into my room. Finally I thought, "It must be the maid," so I cleared my throat loudly and said, "Just a minute."

The door continued to rattle for a few moments more, then abruptly stopped. I thought it was strange, but not strange enough to jump out of bed right away to investigate. My alarm clock was set to go off in a half hour—I couldn't see rushing things.

Ned Seaman was in the room next to mine. We shared a bathroom, sandwiched in between our two rooms. Through my bathroom door I heard Ned shout, "That son of a gun!"

Ned's outburst was followed by the unmistakable sounds of someone hammering on a door. The hammering and shouting went on for another five minutes or so, until I finally yelled, "Ned, what's going on?"

Ned sauntered through the bathroom door and said, "Richards has locked us in our rooms!"

He went to my door and tried to open in, but it was jammed as tightly as his had been.

I asked Ned what was going on again, and he told me the story. While Jim and Dave and I were in K-town the night before, Tommy and Ned had tricked the desk clerk into giving them a spare key to Dave's room. They intended to win their on-going practical joke contest with Dave at all costs.

First, they short-sheeted his bed. Next, they rubbed anti-perspirant onto his toothbrush. They pulled another half dozen little stunts, such as putting tacks inside his flight boots, and then topped it all off by setting Dave's alarm clock to go off at 2 am, and hiding it under his bed!

After discovering his bed short-sheeted, Dave stayed up half the night looking for all of their other little booby-traps. His plan to pay them back was to lock them in their rooms, (which were on the third floor), and then pull the fire alarm.

He used quarters jammed into the crack between the door and the door frame to wedge the doors shut—which explained the rattling I had heard at my door earlier. Dave knew he had to jam my door shut also, or Ned would be able to escape through my room.

The problem turned out to be Tommy's neighbor. Since Dave didn't know who it was, he decided not to pull the fire alarm. He was afraid the guy might jump from his third floor window.

Tommy finally managed to pound on his door hard enough to dislodge the quarters jamming his door. He immediately went to Ned's room and sprung him free.

I finished my shower and left through Ned's room. I don't know if the maid ever did get my door open.

Dave was waiting for me on the landing of the second floor as I made my way downstairs to the office. He apologized for jamming my door, but explained that I was simply an innocent bystander who happened to get in the line of fire between him and the loadmasters.

Ned and Tommy were waiting for us at the foot of the stairs. Dave tried to act nonchalant. He turned to me and said, "Did you hear that girl kicking and scratching and yelling at my door last night?"

I shook my head, and Dave said, "It was awful— I finally had to get up and let her out so I could get some sleep!"

Ester's crew was already waiting downstairs when we got there. They had been waiting for awhile.

The crew management system had changed the way crews obtained their transportation. For decades, whenever the crew wanted a crew bus, the aircraft commander called the motor pool and told them what size bus they needed, where, and when they wanted to be picked up.

Someone decided to change all that. Now, they wanted the desk clerk in the billeting office to make the call for us. They even changed the phone numbers, so that if an aircraft commander tried to call the motor pool directly, the phone rang at the desk in billeting. The clerk would then pick up his direct line to the real motor pool dispatcher, and relay the pertinent information.

You can see what's coming. The desk clerk kept screwing up the transportation order, and Ester and I had no way to fix it. We kept asking for crew buses to pick us up at the billeting office, and the desk clerk kept asking motor pool to send crew buses to some other building!

She kept ordering more buses and motor pool kept sending them, until finally all the buses and drivers were dispersed all over the base at various billeting buildings, with no way to contact the drivers.

Finally, one of the drivers took it upon himself to drive over to the billeting office to see what was going on. Both of our crews jammed onto his bus.

I wanted to go directly to the mess hall, which had received special instructions to stay open longer than usual to accommodate our crews. Ester wanted to go straight to the aircraft. He was nervous about making a late takeoff, even though we were only repositioning the aircraft back to Koksijde. Nobody at Koksijde particularly cared when we showed up.

We took Ester's crew to their airplane, where he left his enlisted crewmembers. They never got breakfast.

We dropped Ester and his officers off at base operations so they could start flight planning. Finally, we had the bus to ourselves, so we went to the mess hall. They had just closed.

I wanted my crew to have breakfast, so I told the driver to take us to the new Burger King just outside the flight line.

After breakfast, we dropped Jim, Ned and Tommy off at the airplane while Dave, Terry and I went to base operations. I don't know what Ester's crew had been doing all that time, but we were able to complete our flight planning chores in short order, and nearly beat Ester back out to the airplanes.

Somehow, in their haste to prepare their aircraft and working with no food in their systems, Ester's crew accidentally jettisoned some fuel out of their wing tanks and onto the ramp. Although they were ready to start before us, they couldn't crank their engines until the fire department washed down the ramp under their ship.

So, the ground handlers walked over to our plane, and we cranked engines and taxied in front of Ester. That meant that we got to take off first, and in the short 1.1 hour flight back to Koksijde, Ester didn't have enough time to pass us.

He was furious when he caught up to us in the operations hooch at Koksijde. We only made him more angry by smiling at his tantrum. After he stormed off, his copilots stole over to us and said, "Is there any way we can transfer to fly with you guys?"

Gary Wilfong met us in operations and confirmed that we were going to be released until Friday. Actually, he had told us that over the telephone when we called him from Ramstein.

One of my loadmasters, Tommy Parsons, had a very good hang-gliding friend in Germany whom he intended to visit if we were granted a couple of days off. Instead of making him ride back to Koksijde with us in an empty airplane and then paying for a train ride back to Germany, Gary agreed to let me release Tommy at Ramstein.

That gave Tommy more time to spend with his hang-gliding buddy in Germany, and saved him some money to boot. The only stipulation was that Tommy had to be in Koksijde on Friday. I knew he would do whatever it took to be there. Guys like Tommy don't take advantage of people who do them favors.

We finally corralled someone from operations and made him drive us back to the Holiday Park in one of the operations vehicles. Nobody knew where the crew buses were.

It was a little late in the day to start out on another adventure, so we unpacked our bags, did our laundry, and planned what we wanted to do for the next couple of days.

All of the other crews were away flying, and everyone on my crew except Dave and I were too tired to be up for any long-range adventures. We decided to take the trolley to Oostende and reserve a rental car for the following morning.

Avis quoted us a price of $54 per day, including

everything except fuel. It was the best deal we could find, so we reserved a car.

We took the trolley back to Koksijde, where we treated ourselves to a long run and a good dinner in the room. We retired early, eager to start our adventure feeling well-rested in the morning.

Chapter 18

"Beep-beep-beep-beep-beep!" The staccato beeping of the altitude alert warning jerked me out of my slumber like a splash of cold water in the face!

I sprang upright in my seat thinking, "Why have we descended off our altitude?"

Wham! My forehead smashed into the low sloping ceiling over my bed, and my head bounced back down onto my pillow.

"Beep-beep-beep-beep-beep." My watch continued beeping away from the lip of my shoes, just a few inches away from my ear.

I heard Dave stifle a laugh into his pillow. When he had control of his voice, he asked, "Excuse me sir, is that the signal to get up, or to smash our heads into the building?"

I mumbled something about dreaming and thinking my watch was an altitude alert warning, but I don't think he heard it. He was too busy laughing at his own joke.

While Dave showered and shaved, I put some water on to boil so we could have a bowl of instant oatmeal and a cup of coffee before we caught the trolley. A half hour later, we were stepping out the door with our small overnight bags slung over our arms. Terry came downstairs just in time to say goodbye, and remind us to call whenever we got where we were going.

The trolley ride up the coast was particularly enjoyable this morning. The sky was clear and the ocean was calm. People were riding horses on the beaches. I slipped my radio out of my bag and placed the headphones over my ears. A sad litany of Irish folk songs drifted over the English channel and into my headphones, enhancing the foreign character of the ride.

In Oostende we picked up our car and a map, although we later decided to stop and purchase a better map. The one that Avis provided only showed major thoroughfares. We had decided to drive to Paris via the beaches of Dunkirk, which were not exactly located on major thoroughfares.

I drove while Dave tried to read the map. I knew it wasn't his fault that he couldn't find his way on the inferior map, but that didn't keep me from harassing him about it.

I kept saying things like, "Newt would know where we are," or "Newt wouldn't have this much trouble."

Newt Honeycutt is undoubtedly one of the best navigators to come along since Magellan. He's one of those guys who has a sixth sense about where he is, whether it's on the ground or in the air. Everyone always likes to fly with Newt, especially me. He's pulled my fat out of the fire more than once.

Dave finally got tired of the harassment and insisted that we stop and buy a better map. He paid for it. I guess we should have split the price—it probably saved us a fortune in gas.

It's a long way to Paris from Koksijde. Even at the speeds that the French drive, it takes a long time to make it to the city of lights.

Dave and I started passing the time telling airline stories. Dave started it off by telling about a recent lay over he had in Gainesville, Florida.

Dave's first officer on the trip, a Texan named Bill, was fond of whistling. He whistled several tunes while they waited for the hotel van to pick them up at the airport for their short overnight stay.

As the van driver drove them to the hotel, he warned them that they were having a dog obedience convention at the hotel. He said that several of the guests got up very early in the morning to walk their dogs, and he warned them to watch where they stepped as they made their way to the office in the morning.

They thanked the driver for the warning and then promptly forgot about it until they reached the hotel. They were walking down the hallway looking for their rooms, when Bill felt an overpowering urge to start whistling again.

Unconsciously, Bill started whistling a fine Texas tune, "The Yellow Rose of Texas." Everything was fine until he reached that high note that corresponds with the lyrics, "The only girl I *love*."

That *"love"* note was exactly the right pitch to set

off the dogs. First, the dog behind the door that Bill was walking past started a loud, "Howoooool" — trying to harmonize with Bill.

The howl was immediately picked up by the dog in the next room, and so on like dominoes until the entire hotel was filled with the howls of singing dogs!

Dave's crew hurriedly found their rooms and locked themselves inside before they could be blamed for the fiasco.

The following morning, two particularly ostentatious ladies with New York accents were waiting in the lobby when Dave's crew showed up. He soon learned that they were going to be passengers on his flight to Charlotte, where they would connect with another flight to New York.

The ladies smoked cigarettes and complained in their high-pitched nasal voices about the inferior quality of everything in the hotel—especially the soundproofing. They were particularly upset about the noise that all the dogs had made last night at midnight.

Bill sat quietly next to the van's side door during the ride to the airport. He wanted to be the first one off the van and away from the ladies.

A light rain was falling so the van driver parked as close as he could to the curb. The driver hopped out and opened the side door, then placed a small step on the street so that Bill could step out of the high van, onto the step, and then onto the curb.

Bill felt like the driver had placed the step just a little too far from the van—he had to stretch his leg to reach the step. Wanting to be helpful, Bill decided to turn around and kick the step a little closer to the van for the New York lady exiting behind him. He would have moved it by hand, but it was sitting in a stream of running water off the curb.

Unfortunately, Bill kicked the step just a little too hard. On the wet street, it slid more easily than he expected.

The first lady had already started stepping out of the van, when she saw Bill kick the step up under the van! Unable to catch herself, she took one giant step straight into the gutter.

As she rolled over in the water to get up, Dave said, "I don't suppose this would be a good time to tell her about you and the dogs, would it Bill?"

I told Dave a story that was told to me by Ken Hyde at Oshkosh a couple of years ago. Ken is a captain for American airlines, you may have seen him in a National Geographic episode called, "Jennys to Jets." Ken has restored a beautiful WW-I Jenny.

Ken and I were talking about pilot practical jokes one day, and Ken started talking about jokes that pilots would pull on non-pilots.

His favorite example was a Louisiana crop duster that he had known named Mr. Walker.

Mr. Walker hired a large field hand each year

during crop dusting season, to do the heavy work of filling the plane's hopper. Often times they would be working on several small fields located close to each other, and Mr. Walker would force his hired hand to actually lie down on the lower wing of his one-man biplane, and hang on while he "hopped" over to the adjacent fields.

Walker always kept the airspeed low for these hops, but he would make a point of flying as close to the trees as he could, taking delight in the terror he saw on his employee's face as they barely brushed past the tree tops.

One day, the engine quit just after takeoff on one of Walker's "hops", and he was forced to put the plane down straight ahead in a swamp.

The wings ripped off when they smashed into the trees, separating the loader from Walker, who wound up trapped in his cockpit in the swamp.

The loader disentangled himself from the smashed fabric of the wings, then made his way to Walker's half-submerged cockpit.

When Walker spotted his loader standing over him he pleaded, "Get me out of here!"

The loader just shook his head and said, "I swear Mr. Walker.. you'd do damn near anything to scare me, wouldn't you?"

The Paris skyline finally appeared on the horizon, so we tried to button down and figure out exactly where we wanted to go and how we were going to get

170

there. It was a futile effort.

Even if we'd known where we wanted to go, we wouldn't have been able to get there among the crazy, horn-blowing, fist-shaking, fender-crunching Parisians. Madrid is the only city I've ever been in where I saw traffic that rivaled Paris—and I had enough sense not to drive in Madrid.

We eventually made our way downtown and parked the car. We decided we would be safer on foot.

After trying about a dozen hotels, we finally found a room in a little dive called the "Modern" Hotel. It wasn't. Actually it was barely more than a broom closet facing an alley with two tiny cots and a dirty bathroom, but it sufficed for our needs. We didn't go to Paris to hang out in the room.

The desk clerk in the Modern was the only Parisian we met during our entire stay who was actually nice to us. The rest were either deliberately rude, or made a point of ignoring us.

We followed the desk clerk's instructions about riding the subway, getting off at the stop that she recommended instead of the one that we would have chosen. We were glad that we did it her way. When we emerged from the underground, we were treated to a spectacular view of the Eiffel tower.

We took some pictures and strolled the length of the mall. We were going to eat at the restaurant in the Eiffel tower, but decided instead to stroll back across the Seine river and eat at a sidewalk cafe. We found a great cafe just north of the Grand Palace. The food was great, even if the waiter was a pain.

The stroll back to the Modern hotel was enough to wear both of us out. Our upstairs room was quite warm, so we opened the windows facing the alley.

The last thing Dave said was, "Be sure to turn off the alarm on your watch... I hate to watch you slam your head into the ceiling."

I fell asleep listening to the sounds of Paris wafting through the curtains, and Dave's giggling.

Chapter 19

The Parisian moonlight finally gave way to the sun. Sunlight sliced through the paper-thin curtains and lit up our room like a photographer's studio. We were up and about early.

We wandered down to the sidewalk cafe on the corner for French coffees and souffles. It was the first time I'd had a souffle for breakfast—it was great.

I first had French coffee in a Vietnamese restaurant in Boston. It's rich stuff. They pour about a half-inch of canned milk in the bottom of a glass, then fill the glass with coffee. You stir the concoction together until it resembles a chocolate pudding, then drink it. It's rich. We used to drink them in the winter in Boston because the hot glass would warm our hands.

After breakfast we checked out of the Modern hotel, saying good-bye to the nice desk clerk, and headed for the Arch de Triumph for one last picture opportunity before heading home.

We were lucky enough to find a parking spot within a block of the Arch. We snapped our best pictures before we even left the parking spot. We took turns posing in the street with the Arch behind us, while the other knelt with his back to the oncoming traffic and snapped the picture. When the "subject" sprinted out of the viewfinder, the "photographer" knew it was time to get out of the street!

Try as we might, we couldn't find a way to get across the famous circle of traffic and actually reach the arch. We could see people under the arch, so we knew there must be a way, but we couldn't find it. We tried asking several Parisians, with the usual results. Dave finally succumbed to the old joke and asked the last guy, "Can you tell us how to reach the area under the arch, or should I just go make love to myself?"

I believe that somebody told us that you have to take the subway to the arch. At least that's what we understood him to say. We were in no mood to go find a subway station.

So, with a hearty, "Follow me," Dave led the charge across the traffic circle to the Arch de Triumph. I'm sure the Parisians are still talking about it.

It was May 13th, 1992. The traffic looked relatively light, so we started our dash. There must be about twelve lanes of traffic circling the arch, which the French expand into about thirty. We were about halfway across when the onslaught came.

It reminded me a lot of those stampede scenes in old westerns. The cowboy is caught afoot, with about a thousand head of cattle headed straight for him.

It was too late to go back, and too late to go forward, and there was nowhere to stand in the middle. There were no lines to stand on for refuge.

We could hear the excited shouts of the people on the sidewalk behind us. The event rapidly took on a sporting air from their perspective.

Dave and I were separated almost immediately. We gave up on trying to go across the circle, concentrating instead on skipping laterally left and right to stay between the constantly undulating streams of cars and motorcycles.

The pedestrians on the sidewalk thought we were goners for sure a couple of times, but a quick dodge under a handlebar or a high-jump over a fender kept us in play—much like a steel ball in a pinball machine.

Incredibly, Dave and I wound up in the same spot at the same time, just one car-width away from safety. There was no time to talk about our next move. A truck was bearing down on us.

We waited until a tail-gating sports car slipped by so close that his mirror brushed both of our legs, then we both leaped for the safety of the center island.

The truck rushed by inches behind us with his horn blaring and laughter spilling from the cab.

Our only consolation was the wild applause of the people back on the sidewalk. They had traveled about a quarter of the way around the circle to keep us in sight as the traffic carried us downstream, and now they laughed and clapped and shouted their appreciation for the show.

Like I said, I'll bet those pedestrians are still talking about that.

We shot up the rest of our film taking pictures under the arch. We sat in the sun for awhile and watched the traffic and tried to talk to the French policemen on duty under the arch. The only real information that we got out of them was that the names inscribed on the walls of the arch were all French generals. They also pointed out the way to the underground walkway back across the traffic circle.

They told us not to try to cross the traffic circle again—as if someone could pay us enough to get us to try it again.

We were only too happy to use the underground to return to the other side of the circle. It came up about a hundred feet from where we had given up on finding it earlier.

Dave drove while I tried to navigate my way out of Paris. Now it was Dave's turn to harass me about not having Newt along. He managed to drive around the Arch de Triumph's traffic circle two or three times before we left the city. I think that being in the circle again with a vehicle of his own was catharsis for him.

We finally broke away from the city traffic and got our clunker up to combat speed on highway A-1, the road leading north back to Belgium.

Several miles whisked by our open windows as we contemplated the French countryside in silence. It truly was beautiful country.

I suppose Dave started getting bored, because he decided to try to talk above the noise of the straining engine and the wind whipping through the windows.

Dave asked me if I'd ever heard the story about

Colonel Pete at the MAFFS training session in Idaho.

I told him I hadn't, since Colonel Pete retired from the Guard several years before I joined the unit.

Dave proceeded to tell me the story of Colonel Pete's tour as the detachment commander at MAFFS training back in 1975.

MAFFS is an acronym which stands for Modular Airborne Fire Fighting System. It's a self-contained fire fighting system which can be installed in the cargo compartment of a C-130 to turn the airplane into a water bomber to fight forest fires.

The sequential system consists of five huge 500-gallon stainless steel spheres which are filled with an orange-tinted fire extinguishing agent. The agent is mostly water, but it does contain other components such as fertilizer.

In addition to the huge spheres containing the fire extinguishing agent, the system also contains five smaller spheres which are filled with compressed air.

The compressed air is used to force the agent out of the spheres, and into two big tubes which are mounted along both sides of the unit. These tubes are eighteen inches in diameter, and they run the entire length of the MAFFS system, from the forward-most sphere all the way out the back of the airplane.

When fighting fires, the copilot presses a "drop button" in the cockpit when the airplane is over the desired point. Pushing the drop button opens valves on each sphere, which allows the compressed air to enter the sphere, force the agent out into the drop tubes, and then all the way out the rear of the airplane.

The agent exits the drop tubes in two giant streams. These water-injected jet streams give the airplane a terrific hydraulic-jet boost for about nine seconds.

Back in the old days, crewmembers were actually known to put on a restraining harness, then slide outside the airplane astraddle one of the drop tubes.

I've heard that it was a fantastic ride, sitting backwards about a hundred feet above the trees with nothing to hold onto except your legs squeezing the tube.

The drops are always made going downhill, usually down a steep ravine. What a terrific rush that must have been to suddenly feel the airplane pitch over at the top of the ridge line, then feel tons of water and agent gushing out of the tube while the airplane accelerated ever faster down the ravine. Then, when the agent was expended, the steep climbing turn to escape the ravine while you tried to hang on and keep your mind off the trees racing by a hundred feet below at 150 knots! Now that *must* have qualified for an E ticket.

After the airplane completed it's escape, the daring crewmember would normally climb back into the airplane, remove the restraining harness, then sit quietly while the airplane sortied back to the the refuel area to take on more agent and compressed air.

Well, the way Dave tells the story, this particular

year the training cycle included one particularly crazy crew from the western U.S. which was famous for riding the tubes all the way back to the field. They came to be known as the "Crazy Eights."

About midway through the training cycle that year, a general from one of the Guard units decided to pay the field a visit and see just how well his men were doing.

Colonel Pete obliged the visiting general with a tour of the training facilities, then offered to drive him out to one of the drop areas so he could watch a drop from the ground.

The drop areas were several miles from the training facilities in Spokane. About halfway to the drop areas, Pete could tell that the general was already tired of riding in the car.

They were just coming up to Lake Coeur d'Alene, where a beautiful resort had recently been completed on the lake's north shore. Pete suggested that they stop at the resort to take on a refreshment before continuing the long drive up into the mountains. The general readily agreed.

Pete and the general ordered drinks from the resort's bar, then settled into a couple of seats outside on the veranda. They were just enjoying their first sips, when the quiet lake scene was suddenly shattered by the unmistakable sound of a C-130 flying at low level. *Real* low level.

Pete spun around in his chair just in time to see the

Crazy Eights go roaring by—flying so low that they were blowing over small sailboats with their prop wash!

To make matters even worse, after the airplane passed, they were flabbergasted to see two crew-members sitting astraddle the tubes protruding from the back of the airplane, waving at the capsized sailors!

The general was so outraged that he couldn't speak. He would have gladly court-martialed the entire crew if he could have gotten his hands on them immediately, but Pete convinced him that by the time they drove back to Spokane that all the crews would be released and back in their quarters for the evening.

The general said that he would handle it first thing the following morning then—immediately after Pete's briefing.

Pete hoped that the general would cool off a little overnight, but the following morning he seemed to be just as upset.

Pete called the briefing to order with a time hack, then started going through the standard briefing items for the day's training events: airplane assignments, training drop areas, etc.

It was about five minutes prior to the prearranged weather briefing time, when the telephone on Pete's podium suddenly rang. Pete customarily answered the telephone during his briefing by turning on the speaker phone. This way, all the crews could copy the weather briefing at once. Pete frowned when the phone rang, assuming the weather briefer was calling early.

But, he couldn't continue his briefing with the phone ringing, so Pete said, "We'll interrupt the briefing at this time to receive our weather briefing. Immediately following the weather briefing we'll complete our formal briefing, then the general will be speaking to us."

Pete shot a hard glare toward the crew of Crazy Eight as he said the last part.

The phone rang again, so Pete turned his attention away from the Crazy Eight crew, and turned on the speaker phone.

After turning the speaker volume up so that everyone could easily hear the conversation, Pete said, "Briefing room."

To his surprise, instead of the voice of the young lieutenant who normally briefed the weather, the room heard the wiry voice of an Idaho farmer ask, "Who is this?"

Pete replied, "This is Colonel Petersen of the North Carolina Air National Guard. Who is this?"

The voice answered, "Are yall the ones flying these here airplanes real low up over the mountains north of lake Coeur d' Alene?"

Pete said, "Well sir, my crews are performing airborne fire fighting training over the designated drop areas which are north of lake Coeur d' Alene.

Their training does require them to fly very low while they are dropping their loads—as low as a hundred feet above the trees."

Obviously unimpressed, the farmer retorted, "Well now you listen here—I got me a pregnant cow up here!"

Pete finally broke the awkward silence that ensued by leaning close to the speaker, clearing his throat, and saying, "Well sir, my boys have only been out here for less than a week—I don't think any of them did it!"

The general was still too overcome with humor to conduct a good butt-chewing after Pete's briefing. Pete promised to take care of it, and the general let him. That was the last of the Crazy Eight crew's tube-riding antics.

By the time Dave finished telling me the story we were almost out of gas. We stopped at a roadside filling station/store to refuel. The gasoline cost as much per *liter* as it did per *gallon* in the U.S.

Much to our surprise, the proprietors of the business were as kind and helpful as they could possibly be. It was such a dramatic difference from the way we had been treated in Paris, that we almost didn't know how to take it. We wound up staying and having lunch at their food counter. No lunch counter in any part of the southern United States could have displayed as much hospitality and good will.

When we finally pulled back onto the highway

and started accelerating back to combat speed, Dave said, "It sort of restores your faith in mankind, doesn't it. I wouldn't mind coming back to France again some day."

"France yes," I nodded, "but I think I've had enough of Paris for awhile."

"Couldn't agree more with you buddy," Dave said as he slammed the shift lever into fourth and buried the gas pedal, "I couldn't agree more."

Chapter 20

May the 14th, 1992, dawned clear and bright. I was awake with the first distant rays of dawn. The moment I awoke, I knew that today would be a special day. Like my wedding day, or the days that I pinned on my wings for the very first time in the Army and the Air Force—I knew that today would be a very special day.

I had offered to let anyone come along with me today that wanted to, but everyone had other plans. Actually that was just fine with me. I had scarcely had five minutes alone for the past two weeks. I relished the idea of spending a day alone with my thoughts.

Although, I knew that with no radio in the car, the drive all the way to Margraten, Netherlands and back would get to be a little tiresome. I've flown across Belgium so many times that the country doesn't seem to be very large, and indeed, it's not.

But, driving across the entire country from its coast to its eastern border is about like driving across Kentucky from west to east—it's a lot further than you think.

There was another reason that I wanted to arrive in Margraten early in the day. I wanted to see if I could find the Dutch sisters who cared for Eugene's grave from 1945 until his body was reinterred to the U.S. in 1949.

Eugene's family lost touch with the sisters after Eugene was reinterred. They had always corresponded with Eugene's mother in Latin—but she died 30 days after Eugene was brought home. All of the old letters and envelopes from the sisters disappeared.

The only thing I had to go on was an old black and white photograph of Eugene standing next to one of the sisters. On the back of the photograph was the name: Amy Reimard.

Amy would undoubtedly be in her late 60s now, or perhaps even her 70s. Although I bear some resemblance to Eugene, she might not remember him. She might well have passed away by now. But, maybe I could find her family. Maybe her sister was still alive, or if she had children, maybe I could find them. They might remember hearing her speak of Uncle Eugene.

I wolfed down some instant oatmeal so I wouldn't have to stop for breakfast on the way, then fixed a large cup of coffee to go and hit the road.

I followed the coast up to Oostende, then turned east on E40. I followed E40 past Brugge, Gent and Brussels, all the way to a little town called Zetrud-Lumay. There I turned north on highway 25 to Tienen, then east again on highways 24 and 32, past Tienen, Sint-Truiden, and Tongeren to Maastricht.

Especially after I turned onto the smaller roads, I started wishing that I had Newt Honeycutt with me again to navigate. The problem was the road signs.

On the road map, every town was printed in two languages, Flemish and French. Here is a short comparison: Brugge (Bruges), Gent (Gand), Brussel (Bruxelles), Tienen (Tirlemont), Sint-Truiden (St. Trond), Tongeren (Tongres).

Unfortunately, the road signs often only displayed one spelling—and I could never be certain as to which it would be. I was trying to make the best time I could on the two-lane roads, looking for the turns that led to the next town along my route. Everytime I decided to be on the look out for sign reading Tienen, I would flash by one that read Tirlemont, or vice versa. I definitely could have used a full-time map reader, but I made it.

When I finally rolled into Maastricht, one of the first things I came to was a sidewalk cafe beside a park. An elderly lady was sitting at a table alone, so I decided to stop and ask for directions to the cemetery.

If you remember what I wrote about our earlier experiences in Amsterdam, you will quickly realize why I decided to ask an elderly person for help. Maastricht is just across the border, in the Netherlands. The last time we went to the Netherlands Brendan's camera was stolen in the first five minutes of our stay.

The elderly lady was very nice, but she didn't speak much English. She did comprehend that I was looking for the cemetery at Margraten. She took the map from my hand and pointed to the road east, then said, "Aachen—ten kilometers," indicating that I should follow the road signs to the Dutch town of Aachen for ten kilometers.

I thanked her, then climbed back into the car and started following the signs to Aachen. Just as I was about to drive out of Maastricht, I passed a huge building with POLICE stenciled on the outside.

"They might help with my search for Amy," I thought, so I parked the car and went inside. I was the only person in the building that I could see.

A female officer stepped out of an office and started past me in the hallway. I said, "Pardon me, but I was wondering..."

She cut me off abruptly with a gesture toward the other end of the hall and a curt, "Desk sergeant."

I almost turned around and left the building. It was not a very hospitable greeting. But, then I reasoned that she might not speak very much English, and she was just saving both of us some time. I decided to at least speak with the desk sergeant.

I was glad that I did. He was a young guy, and seemed eager to help. He listened to my story, then examined the picture that I offered of Uncle Eugene and Amy Reimard.

After a minute he shook his head and said, "What is the most common last name in America?"

I shrugged and said, "I'm not sure, it used to be

Smith."

He nodded and said, "In the Netherlands, it's Reimard, or a close derivation of Reimard. This lady will be extremely difficult to find, *if* she is still alive."

He checked the files that were available to him, then shook his head and said, "Nothing there."

Then, he pulled the phone book for Maastricht and its neighboring communities out from under his desk, and turned to the R section.

As he scanned the pages he said, "The spelling on the back of your picture is peculiar. Usually this name is spelled with the 'i' before the 'e', and the name usually ends with a 't' instead of a 'd'. If we can find an Amy Reimard in this phone book with exactly the same spelling as we see on your picture, then it might be her."

But, the phone book had no such listing.

The sergeant said, "You could always try running an ad in the newspaper, but I'm afraid that you would stand very little chance of success. If this lady or her family lived in this community where they would read the newspaper, they would undoubtedly have a telephone listing."

He could read the disappointment in my eyes. I had briefly related the connection of Eugene and Amy, and I'm sure that he wanted to send me off with some hope.

So, he said, "This phone listing does not include the actual village of Margraten. That village is ten kilometers away on the Aachen highway. You can always check the phone book listings when you get to the cemetery."

I smiled appreciatively, and the sergeant met my eyes as he returned the smile. Then, he did something that surprised me. He reached across the counter and gripped my arm—a reassuring grip that you offer a friend or loved one.

His eyes were still looking into mine, but his had a faint mist spreading over them. I had the definite feeling that he had a story to tell—but I also knew that he wasn't going to tell it.

A second later he released my arm, and grasped my hand to shake it. I thanked him again, picked up the photograph, and walked away.

The drive to the cemetery was very beautiful, lush green rolling hills that rise ever higher as they stretch east away from the Maas river valley and its city. The Aachen highway is actually an old Roman road. It was used by Napoleon in his campaigns in the region.

I turned off the highway at the American Cemetery sign, and followed the roadway back into the hills to the parking lot. I made sure to take my camera and the Amy Reimard photograph, then I locked up the car and walked around to the front entrance.

The first thing I saw was a tower over 100 feet high. A long reflecting pool stretched from the base of the tower, all the way forward the front entrance steps. On the right side of the steps was the visitor's building, and on the left side was an open marble building called the museum.

I couldn't tell at first that it was supposed to be a museum, it just looked like an open building with *colors* on its interior walls.

"Oh no," I thought, "some delinquents have spray-painted graffiti in that building."

But, as I walked closer, I saw that the colors were actually huge arrows affixed to a giant map engraved on the interior walls. Beside the map, the history of the land war in Europe in WWII was engraved on the walls. The colored arrows showed the movements of the various armies involved in the battles.

I was relieved. The museum was exactly the opposite of a graffiti scarred monument—it was a spotlessly-clean testament to the achievements of the allied armies in WWII.

Actually, everything was immaculate. When I drove in from the highway, I passed several Dutch workers who were carefully trimming the grass around the fruit, oak, hawthorn, and maple trees lining the road. All the trees, hedges and lawns were perfectly groomed.

From the front steps, I couldn't see the grave markers. They were obscured by the giant tower, (which also housed a cathedral), and a tall hedge with several trees which surrounded the cemetery grounds.

I started to walk the length of the reflecting pool so that I could enter the graves area, but I stopped when I came to the beautiful bronze statue of a young woman in flowing gowns with doves flying beside her shoulder. The statue is titled Peace.

It was very beautiful, especially its reflection in the pool. I thought that if just this statue was going to have such an emotional effect on me, that I had better take care of my business before I saw all of the markers.

I walked back to the visitor's building, where I was lucky enough to encounter an American named Daniel Neese. I perused the displays in the visitor's building, then introduced myself to Daniel and explained why I was there. He was very helpful.

Daniel looked at my pictures of Amy and Eugene, and at some others that I had of Eugene's old grave marker at Margraten. He was very impressed by the old photo of Eugene's wooden cross, saying that those pictures were quite rare.

He explained that the cemetery was originally established on 65 acres of farmland, granted in perpetuity, without charge, as a final resting place for American military men killed in WWII. Almost all of the men buried there were killed liberating Holland from the Germans.

Originally the cemetery contained the graves of over 19,000 Americans. Their graves were marked with wooden crosses and Stars of David. Just as my grandfather asked for the body of Eugene to be reinterred, most of the families of the fallen men asked that their bodies be returned to the U.S.

After a few years the number of requests for reinterrment declined dramatically, and the cemetery was redesigned to bring all of the markers back together again.

Today, just over 8,300 American military men are buried at Margraten.

Families still appear at times with thoughts of asking for the reinterrment of a man buried there. Usually the circumstances are, that the parents of the deceased have finally passed away, and now the brothers and sisters want their family member brought home. Almost always, after these family members visit the cemetery and see how immaculately it's kept and how well it's designed—they decide to leave their loved one in the honored company of his comrades in arms.

I asked Daniel if there were any records showing where Uncle Eugene was buried, but he said they had all been destroyed. He also said that all of the old wooden markers had been destroyed. Now the markers are all white granite.

Daniel looked at the picture of Amy Reimard and listened to my story, then he readily checked the Margraten phone book for an identical listing. There was none.

Not wanting to give up though, Daniel sent for his Dutch foreman, a man named Berloth. Daniel explained the story to Berloth, showing him Amy's photograph. Berloth shook his head and said that he could not remember the lady, but he only started working at the cemetery in the 1950s.

Then, he slapped his leg and said that Van Laar would know!

Daniel got an excited look on his face and said, "Yes! Terrific idea! Van Laar will surely know!"

Daniel answered my perplexed look by explaining that the prior foreman of the Dutch grounds keepers, a man named Van Laar, had worked at the cemetery since its inception in WWII. He had lived in the village of Margraten his entire life. If this Amy Reimard was from anywhere around Margraten, Van Laar or his wife would surely know her.

Berloth excitedly suggested that since Van Laar didn't have a telephone, that he take me to see Van Laar.

Daniel said, "Yes, of course. Take the pickup truck!"

I was overwhelmed. These people couldn't do enough to help. Berloth hurried off to collect the truck, shouting instructions to his assistants as he passed.

I offered to drive myself to Van Laar's house so as not to interrupt their work schedule, but Daniel wouldn't hear of it.

It was a lucky thing that Daniel insisted that Berloth drive me. I never would have found the place. And if I did find the house, I couldn't have gotten in.

Berloth stopped the truck in front of a large gate outside what can only be described as the quintessential Dutch country home. It was shaped like a horseshoe, with a beautiful courtyard in the middle. The huge wooden gate across the bottom of the horseshoe prevented entry, however, so if I'd been alone I never would have opened the gate and discovered the courtyard.

Berloth was a friend, so he had no reservations about opening the gate. He tried to explain that the Van Laars were old and probably couldn't hear us knocking on the gate, but his English wasn't terrific and I wasn't certain that I correctly grasped his message.

He led me to a split Dutch door at the rear of the courtyard where he knocked again.

This time, Mrs. Van Laar heard the knock. The upper half of the split door swung open, and I saw a remarkably kind, peaceful face appear in the doorway. I knew in an instant that if I was ever granted a Dutch grandmother, she would be my first choice.

Her face lit up like a beacon when she recognized Berloth. She pulled him to the door and hugged him while they both laughed.

Berloth introduced me in Dutch, and quickly explained my business. When he finished speaking Mrs. Van Laar turned to me and said, "Please bring your picture inside. I will get my glasses and look at it for you."

She opened the lower half of the door, then led us through her Dutch kitchen and motioned toward a table in the living room. Berloth and I seated ourselves and waited for her to return with her glasses.

I asked if her husband was home, and she said, "Yes, he's working in the garden. If I don't recognize the lady we will call him, but if this lady has ever lived close to our village, I will know her."

She took the picture from my hand and carefully scrutinized it's images. After several long moments, she shook her head, then turned the picture over and read the name on the back.

Finally, she handed the picture back to me and said, "She is not from this area. She only visited here to decorate your uncle's grave on veteran's day. Many women do that—especially right after the war."

Daniel had already told me about the Dutch lady's groups who make a pilgrimage to the cemetery each American veteran's day to decorate the graves with flowers.

I told Mrs. Van Laar, "This lady knew my uncle. He helped her after the German's left. She wrote to my grandmother."

"Yes, I see," she said, "then it wasn't a casual picture found in your uncle's effects—she had your family's address."

I nodded, then we sat in silence for a few moments

while Mrs. Van Laar looked far away, at old memories that only she could see.

She came back to us and began asking personal questions about me and my family. My wife, daughter, parents, where we lived, what I did, etc. Her questions weren't intruding, just curious, so I spent as much time as she wished describing my life in the United States.

As I talked I absent-mindedly traced my fingers along the patterns in the white lace of the table cloth. When I finished answering one of her questions, Mrs. Van Laar pointed to my hand and said, "Do you like the table cloth?"

I answered, "It's very nice—is it from Brugge?."

I had heard that there were bargains to be had in fine lace in Brugge, but we hadn't visited the shops there.

She answered, "Yes it is. You take it. You take it to your wife."

For a moment I was flabbergasted, then I quickly recovered and said, "No! I didn't mean that I wanted it—I just meant to say that it was very beautiful... that it decorated the table very nicely."

She said, "Do you like the table?"

I said, "It's very nice."

"You take it," she said, "take it to your wife in America and tell her it is a present from me."

Berloth started laughing at my embarrassment, which caused Mrs. Van Laar and me to begin laughing also. Finally I explained that I was flying home, and I had no room on the plane for such a beautiful table.

Afraid of what she was going to offer next, I shyly asked if Mr. Van Laar could please take a look at the photograph. She smiled again, then nodded and said, "Let's go to the garden."

She led the way outside to the courtyard, where she left Berloth and me, saying, "I'll be back in a minute."

I learned a few minutes later that she wanted a chance to let her husband wash his hands and put on a clean shirt. While we waited, Berloth pointed out the layout of the house—the L-shaped living quarters, and the old barn section that made up the other long side of the horseshoe.

He said that the barns were only used for storage nowadays, or as garages for those people who had cars. There was a beautiful antique water hand-pump in a corner of the courtyard.

There were also wooden shoes hanging by their heels on each side of the widows. Tiny ferns were growing from the holes in the wooden shoes.

Mr. and Mrs. Van Laar appeared a few minutes later, with Mr. Van Laar wearing his fresh shirt and a welcoming smile.

He asked to see the photograph, then studied it closely. He shook his head without speaking, then shuffled the photos to study the one of Uncle Eugene's old wooden marker. He held it a long time.

Finally, he handed the photographs back and said, "That was a long time ago. For the first few years after the war, thousands of people would visit the cemetery on veteran's day. I can not remember this single lady. If it was a long journey for her, she probably stopped coming after your uncle was taken away."

I thanked him, then asked if he would mind if I took his photograph. He laughed and said, "Okay," so I asked Mrs. Van Laar and Berloth if they would also be in the picture.

Mrs. Van Laar adamantly refused. However, she insisted on being the photographer, and taking a picture of the three of us.

She ushered Mr. Van Laar, Berloth and I in front of her water pump for the picture. I quickly explained the mechanics of the *disposable* camera to her, and she acted very surprised. She said it was the first disposable camera she had ever seen.

She snapped the picture in front of the hand pump, then I asked if I might have another with the three of us standing in front of one of the windows, framed by the wooden shoes.

"Aha!" she exclaimed. Now she had me.

"You like the shoes?! she asked with a note of disbelief in her voice.

"Yes, we don't have them in South Carolina."

"You take shoes to your wife. Tell her they are

presents from the Van Laars."

She overrode my protests, saying that they were very common in Holland. Mr. Van Laar started to take one off the wall behind us after we snapped the picture, but she insisted on going into their work shop and returning with a freshly-painted shoe. It had a beautiful plant growing inside which reminded me of the small ice plants that I saw growing on the beaches in California.

Mr. Van Laar inscribed their names and the date on the bottom of the shoe. I was very touched.

I was about to say good-bye, when Mrs. Van Laar made a surprising offer. She took me back into the living room to the table, where she wrote down all of the information about Uncle Eugene and Amy Reimard and myself, especially my address.

Then she told me about a book that is published once a year in Holland. To get your story in the book, you must either be trying to contact somebody that you've never met, or someone that you haven't seen in at least twenty-five years.

She said the first half of the book is filled with stories with titles like mine, "Nephew of WWII American soldier trying to find Dutch woman named Amy Reimard." The second half of the book is devoted to stories of people who have actually found each other through the book!

She said that the stories are very popular, especially among people of her generation. She offered to write up my story and send it to the publishers of this book.

The book is printed in Flemish, so Mrs. Van Larr planned to use her address as the contact point. She said that if Amy Reimard was still alive, that she probably read this book, and would get in touch with Mrs. Van Laar after the next edition. Mrs. Van Laar would then put her in touch with me.

It was a wonderful offer, which I readily accepted.

Berloth and I drove away from the Van Laar home, waving back to them. They stood in the open gate waving until we turned a corner and passed out of sight.

Berloth parked the truck back at the cemetery and we went together to Daniel's office to report our findings. He was pleased by Mrs. Van Laar's offer.

I thanked Daniel, then walked outside with Berloth. He met my eyes as he shook my hand, and only nodded as I thanked him. I was glad that I had met him.

As I started walking beside the reflecting pool toward the tower, I couldn't help but reflect on the difference between these kind people and the thieves and drug addicts of Amsterdam. I'm sure that there's something about spending their whole lives caring for the graves of American servicemen that makes them feel kindly toward those servicemens' families. But, beyond their vocations, these kind country people would be recognized as good-hearted folk anyplace.

As I passed by the tower and onto the central mall, I was overcome with emotion. I could not have spoken. Indeed, I don't know the words that could express my feelings now.

A beautiful grassy mall separates the cemetery into 16 plots of graves, 8 plots on each side of the mall. Each plot contains over 500 markers.

The markers are arranged in a fashion which I have never seen before or since. They appear to be lined up straight when viewed from head-on, but they also appear to curve in symmetrical lines when viewed from the side. It's a striking effect.

There are poplar trees planted in the grassy areas between the plots. Fitting, I thought, since Uncle Eugene was from a town named after a grove of poplar trees growing on a bluff above the Little Black river—Poplar Bluff, Missouri.

The mall extends from the tower at the front end of the cemetery to a granite-based flag pole at the other end.

I left the mall to walk along the grassy paths between the graves, where I could read the markers' inscriptions. All the men are intermingled—there is no order such as last name, rank, religion or branch of service. One hundred and seventy-nine of the markers are fashioned in the Star of David—the rest are Latin crosses.

As I walked, I looked up and noticed that from that position, a freshly pruned poplar tree was very neatly aligned with the American flag at the far end of the cemetery. The sun shone through the freshly trimmed branches, dappling the markers before me in splotches of soft light reflecting from the white stone, and soft undulating shadows.

I knelt before the cross in front of me and read its inscription. It read:

HERE RESTS IN HONORED GLORY
A COMRADE IN ARMS
KNOWN BUT TO GOD

I raised the camera to my eye and snapped the last picture on the roll. It is the best photograph I've ever taken.

I said a prayer, then stood again and started walking the grassy path to the flag. I passed more markers to unknown soldiers—there are 106 in all.

There is even one grave with two bodies in it. The men's bodies were mangled together in their fox hole by an artillery round.

There are four posthumous medal of honor winners buried in the cemetery.

I left the grassy paths to walk across the mall to the flag pole. A light wind rustled the stars and stripes over my head. The cemetery was completely quiet except for that sound, I was the only living soul there.

As I looked out across the pure white markers, the only words that came to me were the words of a poem written by a WWI Canadian soldier, John McCrae, back when Holland and Belgium were known as Flanders.

> In Flanders fields the poppies blow
> Between the crosses, row on row,
> That mark our place; and in the sky
> The larks, still bravely singing, fly
> Scarce heard amid the guns below.
> We are the Dead. Short days ago
> We lived, felt dawn, saw sunset glow,
> Loved and were loved, and now we lie
> In Flanders fields.
> ...To you from failing hands we throw
> The torch; be yours to hold it high.
> If ye break faith with us who die
> We shall not sleep, though poppies grow
> In Flanders fields.

John McCrae was a doctor, soldier, and war casualty.

I stayed there a long time. It was very beautiful, very peaceful. I could easily understand why family members would elect to leave their loved ones interred there.

It was starting to get late. I knew it was time to go.

I paid my last respects and clambered back into the rental car. Not wanting to try to negotiate the secondary roads again, especially in the dark, I decided to take highway E25 south from Maastricht to Liege, and then pickup E40 for the long run all the way back across Belgium.

Even with the accelerator on the floor, the drive took a long time. It was a long time to be alone with my thoughts and emotions.

I kept thinking about the relationship of pasts and futures.

In his terrific book, *The Cannibal Queen*, Stephen Coonts says that the view is always different over the tail than it was through the prop. That obstacle that looked so imposing when you were headed for it, doesn't look so tough when you're looking back over the tail at it.

As I drove through the darkening countryside, I tried to picture what the area must have looked like to Uncle Eugene as he fought his way through it. He probably drove along the same roads, albeit he viewed the territory from the top of a Sherman tank.

But, as I tried to put myself in his shoes, imagine what he must have seen and how he must have felt about it... I started to feel like I knew. I think he looked out at the awful waste and destruction of war, and was sickened by it. I think he looked at the cost in lives and human suffering, and his heart nearly broke at what he saw.

But, I also think that he could look past the present, and see something better. I think he could buoy himself on the camaraderie and accomplishments of his men, and feel the confidence that it would take to rebuild this land.

I think he could look at the burned-out farmhouses and scorched fields, and see the picturesque homes surrounded by green fields that I saw today.

The more I tried to look back at what Eugene must have seen and felt, the more I could feel him looking forward to today.

When I was growing up, we had two mirrors facing each other in the bathroom of my parent's house. My mother and sisters could look forward into one mirror, and see the reflection of the back of their heads for fixing their hair.

But, if you looked beyond that reflection, you could also see a reflection of your face being bounced back from the mirror behind you. The image kept bouncing back and forth between the two mirrors, until it seemed like there were a dozen images layered within the mirror.

As the images went deeper and deeper, they became ever less distinct, until you could hardly recognize the face.

That was how I felt—like I was looking way down into the reflections on one side, and seeing somebody else looking way up from the other.

I must have passed into that part of your brain that doesn't know what time it is for about a hundred miles. The next time I remember knowing where I was—I was coming to the end of the highway in

Oostende. Time to start paying attention.

I coasted down the seaside road to Koksijde where I found a party taking place in our hooch at the Holiday Park.

All the crews were back by now, and they all had tales to tell. Todd and Ray had spent some time landing on dirt strips in what used to be East Germany.

Todd told us about reading some graffiti printed above a urinal in a men's room in the old East Germany. The only graffiti printed in English said, "Moose and squirrel must die! Boris."

Brendan and Pick had experienced some hair-raising times in England, while Artis and JT had enjoyed themselves in the local area. I briefly related my day's adventure—everyone liked the wooden shoe. It was a great little reunion.

The crowd finally thinned out so we turned out the lights and retired to our attic abode. Dave commented that I seemed to have enjoyed my day immensely.

I agreed that I had. It was a very special day that I had waited a long time for.

It was worth the wait.

Chapter 21

The next three days rushed by in a blur. Everyone was doing whatever it took to get us homeward bound.

Friday was consumed with the tasks of packing and loading up the airplanes. The magnificent seven got together for one last hurrah in Oostende, which basically consisted of buying Godiva chocolates to take home and feasting on Belgian waffles. Most of the chocolates melted.

We were scheduled to depart Koksijde at one hour intervals on Saturday for the first leg of our flight home—the leg to Lajes. When each crew showed up we were told that all of our flight plans were already filed, that all we had to do was make an on-time takeoff.

That didn't work out exactly as planned. I was flying the third ship out. When we got to the Land's End VOR over the English coast, we found the two C-130s that had departed ahead of us in a holding pattern there.

It seems that the oceanic portions of our flight plans never got filed, and the oceanic controllers would not give us permission to enter their airspace.

Our B-models certainly didn't have the extra fuel that holding over Land's End would consume, so we declared MARSA (Military Accepts Responsibility for Separation of Aircraft), and kicked out of the holding pattern VFR.

The other ships followed us south until we were finally able to get a clearance to turn west and head for the Azores. We barely made it with legal fuel reserves, but at least we didn't divert into Mildenhall, England with all of our passengers for the night. What a mess that would have been.

The flight from Lajes to Koksijde back on the 3rd of May had taken 5.4 hours. The flight from Koksijde back to Lajes required 6.8 hours!

Everyone got together that night for our "last supper" of the deployment. We had a terrific dinner of grouper and vegetables downtown. The wine, stories and laughter flowed freely.

We were up early the next morning to start the long haul home. We were flying west so the time zone changes were working in our favor, but the winds were working against us.

The winds were too strong to strike out direct for Charlotte, so we had to make a fuel stop in St. Johns, Newfoundland. The flight from Lajes to Newfoundland took 5.2 hours.

We weren't on the ground for an entire hour at St. Johns. We stayed just long enough to refuel and file the next leg of our flight plan, then it was on to Charlotte.

That last 5.6 hour leg from St. Johns to Charlotte was both the hardest and the easiest of the trip. It was like the last mile of a marathon, it's hard, but you can see the finish line.

We used our HF radios during this last leg, to put one last little piece of frosting on the cake. We got a phone patch with a local radio station, and persuaded them to play Linda Ronstadt's "Back in the USA" for us as we made our approach into Charlotte.

We were able to tune in the song and play it over the PA system so all the guys in the back could hear it too. It was great. All three of us touched down within a few minutes of each other, but I think it was Artis's crew that actually landed when she was singing the verse, "I just touched down on an international runway..."

Our families were there to greet us as we walked in off the flight line. My daughter won the race to give me the first welcome home hug—it seemed like she had grown another inch while I was away. My wife drove me home, fed me a steak dinner and made sure I slept as long as I wanted to in the morning.

When I did awaken the following morning I was careful not to sit up too fast for fear of bashing my head on the ceiling. It always takes a couple of days to break old habits.

There's an old saying in military aviation that somebody always utters before a crew breaks up after a long trip... it goes, "I guess it's time to leave our loved ones and return to our dependents."

It's certainly true that after eating, sleeping, and living with a crew for a couple of weeks, you can't help but miss those guys for a day or so after you're home. It's great to be home with the family, but the crew is still there for awhile, just below the surface of your thoughts.

Brenda took the wooden shoe I brought her from the Van Laar's and hung it from a post between the tiers of the deck on the back of my house.

I enjoy sitting on the lower tier of my deck in the mornings with a cup of coffee and the newspaper. Often times, when I finish the newspaper, I'll sip my coffee and contemplate that old wooden shoe.

I haven't heard from Mrs. Van Laar yet, but it's still early—the new book probably hasn't been printed in Holland yet. I sent her a brand new disposable camera the week I returned home. I hope she took some good pictures with it.

My Guard unit is scheduled to begin getting brand new C-130s soon. I'll probably be one of the last pilots to transition to the new planes. I fly brand new equipment at USAir—I kind of like flying the old planes at the Guard. They have a lot of character.

Speaking of characters, the magnificent seven and company are still in full swing. Births, deaths, weddings, divorces, mortgages, airline fortunes-and-furloughs, physicals, check rides, Christmas parties and the United gosh-darn Way Campaign... life goes on more-or-less the way it's supposed to for all of us in Charlotte, North Carolina.

I don't even have to water the plant in that wooden

shoe. It just keeps growing greener and healthier all by itself. It looks good on an American deck.

Uncle Eugene would be pleased.

About the Author

Sherman F. Morgan began his aviation career as an Army helicopter pilot. He graduated from Army flight school (as a Warrant Officer) at Ft. Rucker, Alabama on July 7th, 1976. He was stationed at Ft. Ord, California until July, 1979.

After completing his Army tour, Sherman joined the Air Force Reserve at Travis AFB, California, where he flew C-141 Starlifters. While flying C-141s at Travis AFB, Sherman earned his multi-engine A.T.P. at the Travis AFB flying club.

After a year of flying C-141s, Sherman transitioned to the C-5 Galaxy. He flew C-5s at Travis AFB, California, Dover AFB, Delaware, and Kelly AFB, Texas.

In 1989 Sherman joined the C-130 Air National Guard unit at Charlotte, North Carolina, where he serves as a C-130 pilot.

Sherman flies for USAir from its Charlotte, North Carolina hub, and resides in Fort Mill, South Carolina, with his wife of 20 years, and their 13 year old daughter.

Sherman is always soliciting humorous flying stories for his books. If you have one, write it down and mail it to Sherman c/o Pendragon Publishing Co., 1484 Old Tara Lane, Fort Mill, SC 29715.

Old Planes, Young Men and Red Wooden Shoes

Mail/Phone/Fax Orders

If you do not wish to tear this page out of the book, you may copy it. Mark your choices on the copy and mail it to: Pendragon Publishing Co., 1484 Old Tara Ln., Fort Mill, S.C. 29715. For Visa & Mastercard orders, call:
(803) 548-8433 or Fax (803) 548-2084.

Name:_____

Address:_____

City:_____

State & Zip Code:_____

____Check ____Money Order ____Visa ____ Mastercard

Credit card number:_____

Credit card expiration date:_____

                                            QUANTITY

The Aviation Humor of 1987
$5.95 + $1.55 P/H = $6.50
                                      _____

Classic Aviation Humor—Book II
$5.95 + $1.55 P/H = $6.50
                                      _____

Classic Aviation Humor—Book III
$6.95 + $2 P/H = $8.95
                                      _____

Good Sticks
$19.95 + $2 P/H = $21.95
                                      _____

Old Planes, Young Men & Red Wooden Shoes
$10.00 + $2 P/H = $12.00
                                      _____

Please make checks payable to: Pendragon Publishing Co.

Sherman F. Morgan

Mail/Phone/Fax Orders

If you do not wish to tear this page out of the book, you may copy it. Mark your choices on the copy and mail it to: Pendragon Publishing Co., 1484 Old Tara Ln., Fort Mill, S.C. 29715. For Visa & Mastercard orders, call:
(803) 548-8433 or Fax (803) 548-2084.

Name:_____

Address:_____

City:_____

State & Zip Code:_____

____Check ____Money Order ____Visa ____ Mastercard

Credit card number:_____

Credit card expiration date:_____

                                              QUANTITY

The Aviation Humor of 1987
$5.95 + $1.55 P/H = $6.50   _____

Classic Aviation Humor—Book II
$5.95 + $1.55 P/H = $6.50   _____

Classic Aviation Humor—Book III
$6.95 + $2 P/H = $8.95   _____

Good Sticks
$19.95 + $2 P/H = $21.95   _____

Old Planes, Young Men & Red Wooden Shoes
$10.00 + $2 P/H = $12.00   _____

Please make checks payable to: Pendragon Publishing Co.

Sherman F. Morgan

Mail/Phone/Fax Orders

If you do not wish to tear this page out of the book, you may copy it. Mark your choices on the copy and mail it to: Pendragon Publishing Co., 1484 Old Tara Ln., Fort Mill, S.C. 29715. For Visa & Mastercard orders, call:
(803) 548-8433 or Fax (803) 548-2084.

Name:_____

Address:_____

City:_____

State & Zip Code:_____

____Check ____Money Order ____Visa ____ Mastercard

Credit card number:_____

Credit card expiration date:_____

QUANTITY

The Aviation Humor of 1987
$5.95 + $1.55 P/H = $6.50   _____

Classic Aviation Humor—Book II
$5.95 + $1.55 P/H = $6.50   _____

Classic Aviation Humor—Book III
$6.95 + $2 P/H = $8.95   _____

Good Sticks
$19.95 + $2 P/H = $21.95   _____

Old Planes, Young Men & Red Wooden Shoes
$10.00 + $2 P/H = $12.00   _____

Please make checks payable to: Pendragon Publishing Co.

Mail/Phone/Fax Orders

If you do not wish to tear this page out of the book, you may copy it. Mark your choices on the copy and mail it to: Pendragon Publishing Co., 1484 Old Tara Ln., Fort Mill, S.C. 29715. For Visa & Mastercard orders, call:
(803) 548-8433 or Fax (803) 548-2084.

Name:_____

Address:_____

City:_____

State & Zip Code:_____

____Check ____Money Order ____Visa ____ Mastercard

Credit card number:_____

Credit card expiration date:_____

QUANTITY

The Aviation Humor of 1987
$5.95 + $1.55 P/H = $6.50   _____

Classic Aviation Humor—Book II
$5.95 + $1.55 P/H = $6.50   _____

Classic Aviation Humor—Book III
$6.95 + $2 P/H = $8.95   _____

Good Sticks
$19.95 + $2 P/H = $21.95   _____

Old Planes, Young Men & Red Wooden Shoes
$10.00 + $2 P/H = $12.00   _____

Please make checks payable to: Pendragon Publishing Co.

Sherman F. Morgan

Mail/Phone/Fax Orders

If you do not wish to tear this page out of the book, you may copy it. Mark your choices on the copy and mail it to: Pendragon Publishing Co., 1484 Old Tara Ln., Fort Mill, S.C. 29715. For Visa & Mastercard orders, call:
(803) 548-8433 or Fax (803) 548-2084.

Name:_____

Address:_____

City:_____

State & Zip Code:_____

____Check ____Money Order ____Visa ____ Mastercard

Credit card number:_____

Credit card expiration date:_____

	QUANTITY
The Aviation Humor of 1987 $5.95 + $1.55 P/H = $6.50	_____
Classic Aviation Humor—Book II $5.95 + $1.55 P/H = $6.50	_____
Classic Aviation Humor—Book III $6.95 + $2 P/H = $8.95	_____
Good Sticks $19.95 + $2 P/H = $21.95	_____
Old Planes, Young Men & Red Wooden Shoes $10.00 + $2 P/H = $12.00	_____

Please make checks payable to: Pendragon Publishing Co.